IN
THEIR
OWN
WORDS

Conversations with the astronauts and
men who led America's journey into
space and to the moon.

Edited by Scott Sacknoff

Space Publications LLC
June 2003

In Their Own Words
Edited by: Scott Sacknoff

From the pages of *Quest: The History of Spaceflight Quarterly*
For more information on Quest, please visit:
http://www.spacebusiness.com/quest

Published by:
Space Publications LLC
P.O. Box 5752 Bethesda, MD 20824-5752
http://www.spacebusiness.com

Copyright (c) 2003 by Space Publications LLC
Printed in the United States of America

Cover Photo Credit: NASA. John Young, pilot of the Gemini III spaceflight,
checks his helmet with Al Rochford during suiting operations prior to flight.

ISBN: 1-887022-11-2

Library of Congress Cataloging-in-Publication Data

In their own words: conversations with the astronauts and men who led
America's journey into space and to the moon / compiled & edited by
Scott Sacknoff

Includes index.
  ISBN 1-887-022-11-2 (Trade Paper: alk. paper)
  1. Astronauts -- United States -- Interviews. 2. Astronautics -- United
  States. 3. Outer space--Exploration. I. Sacknoff, Scott. II. Title

TL789.85A1T44 2003
629.45'4'0922--dc21

10 9   8   7   6   5   4   3   2   1

Table of Contents

Acknowledgements

This book represents a collection of interviews and articles that have appeared within the pages of *Quest: The History of Spaceflight.* Published since 1992, it is the only journal to exclusively focus on preserving the history of the spaceflight.

This publication would not be possible without the efforts of the following individuals:

- Professor Stephen Johnson, Suezette Bieri, and Kathy Borgen of the University of North Dakota's Department of Space Studies who edit and produce the quarterly journal;

- Glen Swanson, the founder and managing editor of *Quest* from 1992 to 1996, whose interviews of Jim Lovell and Jack Lousma appear in chapters Six & Eight, respectively;

- Don Pealer, whose interviews of Pete Conrad, Fred Haise, Guenter Wendt and James McDivitt appear in Chapters Three, Seven, Nine and Eleven, respectively;

- Roy Neal, whose interview of Alan Shepard appears in Chapter Two;

- Randy Attwood and Keith Wilson, whose interview of Brigadier General Charles Duke appears in Chapter Four;

- Rick Boos, whose interview of Scott Grissom appears in Chapter Five; and

- Carol Butler, whose interview of General Bernard Schriever appears in Chapter Ten.

Introduction

Eleven years after the Wright Brothers made their historic airplane flight in Kitty Hawk, North Carolina, Robert Goddard was issued his first patents for a rocket apparatus. By 1926, the "Father of American Rocketry" would launch the first liquid propellant rocket. Over the next two decades his research and flight experiments would form the core to a critical base of knowledge that would eventually lead us to the launch of the first orbital satellites and within sixty years to a man stepping on the surface of the Moon. Now, almost fifty years after the "Space Age" began, we can see a vibrant space program that routinely launches satellites and scientific probes that enable a range of applications for commerce, communications, defense, and environmental analysis.

For many, the public interest in space continues to lie in the awe and inspiration that spaceflight brings. In talking with people over the years, you can hear in their voices, the adrenaline rush as they describe their feeling when man first landed on the moon the thrill of watching the launch of a Space Shuttle. It's as if they have put themselves in the pilot's seat as the roar of the engines thrust their vehicle up into the weightlessness of space.

To help preserve the stories behind the research, the engineering feats, and the space milestones, is a small journal entitled, *Quest: The History of Spaceflight*. Published on a quarterly basis since 1992, *Quest* has been offering stories, analysis, and first-person accounts that bring a behind-the-scenes look at the development of programs, policies, and people that helped shape the space industry. *"In Their Own Words"* represents a collection of several of the first person interviews that have appeared over the years.

In reading these stories, what comes across is not just the thrill of spaceflight but the comraderie that existed between the astronauts and the men on the ground. You never hear them talk about 'me' but 'us' and that spaceflight would never be possible without the engineers, the scientists, the managers, and the support personnel -- hundreds of thousands of people all working toward a common goal.

As you read these interviews and stories, it is my hope that you will gain an insight, not only into the space program, but into the people behind it. The true human element of space.

As an example, listen to these words from Robert Goddard,

> "A second and a half later there was a terrific explosion
> with smoke and flame bursting from the bottom of the
> rocket. Undoubtedly, those new to rocket experimentation
> feared that something had gone wrong. But it was a planned
> explosion..."

Even after my own almost two decades in the space industry, there are still moments when I read a story or hear someone tell a personal recollection, that my attention is captured.

For instance, hearing Jim Lovell recount the Apollo 13 mission,

> "At the time the explosion occurred, I looked at my compan-
> ions and said, 'This couldn't have happened at a worse time.'
> We were some 200,000 miles from Earth and going in the
> wrong direction. But I was wrong..."

The twelve chapters inside this book were chosen to provide the reader with a range of insights. It starts with Robert Goddard describing a test flight of one of his rockets as well as his vision for its future use.

To understand the early space program, there is an interview with space pioneer Alan Shepard, astronaut Jim McDivitt and moonwalkers, Pete Conrad and Charles Duke. The Apollo 13 mission is covered by discussions with mission commander Jim Lovell, lunar module pilot Fred Haise and CapCom Jack Lousma.

To get a different perspective, an interview with Scott Grissom, whose father, Gus Grissom, perished in the Apollo 1 fire, is included. Guenter Wendt, known as "The Pad Fuhrer" for his role in the Gemini and Apollo programs, shares with us a number of stories including practical jokes by and to the astronauts.

The military's early role in space is discussed with Gen. Bernard Schriever, who has been called the "Father of the Air Force Space Program."

And lastly, there is a fascinating discussion between Walter Cronkite and his colleagues at CBS, NBC and NASA as they discuss how the media covered the space industry in the 1960s and share some personal stories that are sure to bring a smile.

So whether you are reading this book because of the awe and inspiration that spaceflight brings to so many or for its educational value [or both]; I hope that you enjoy these stories as I have.

May our future include the Earth below us and the stars above.

Scott Sacknoff
Publisher
Quest: The History of Spaceflight

For More Information on
Quest: The History of Spaceflight, **please visit:**

http://www.spacebusiness.com/quest

"Preserving the history of spaceflight...one story at a time."

Acronyms and Abbreviations

A204 - Also known as Apollo 1
AAP - Apollo Applications Program
ALSEP - Apollo Lunar Surface Experiment Package
ARDC - Air Research & Development Command
ARPA - Advanced Research Projects Agency
ARPS - Aerospace Research Pilot School
ASTP - Apollo Soyuz Test Project
CAPCOM - Capsule Communicator
CSM - Command and Service Module
EMU - Extravehicular Mobility Unit (spacesuit)
EVA - Extravehicular Activity
LLTV - Lunar Landing Training Vehicle
LRV - Lunar Rover Vehicle
LM - Lunar Module
LMP - Lunar Module Pilot
MOL - Manned Orbiting Laboratory
MR-4 - Mercury Rocket 4
NACA - National Advisory Committee for Aeronautics
NASA - National Aeronautics & Space Administration
Shuttle ALT - Approach and Landing Test Program
Shuttle OFT-2 - Shuttle Orbital Flight Test 2
SIMBAY - Spacecraft Scientific Instrument Module Bay

Commonly Used Nicknames

Deke - Deke Slayton
Gordo - Gordon Cooper

Robert Goddard: Timeline

1882: Born October 5, 1882

1912: Explored mathematically the practicality of using rocket power to reach high altitudes and escape velocity

1914: In July, Goddard received his first patents for rocket apparatus

1919: "A Method for Reaching Extreme Altitudes" is published by the Smithsonian Institution

1926: On March 16th, Goddard first launches a liquid propellant rocket.

The Past Revisited
by Robert Goddard

The Founder of American Rocket Research

On an April morning in 1940, I stood in the doorway of a small shack at an isolated spot on the high-plains of eastern New Mexico, finger poised above a group of telegraph keys mounted on a shelf beside me.

Pressing two of those keys in proper sequence would send approximately 500 pounds of metal and fuel roaring into the sky; for the electric impulses would launch a rocket which, so far as I can find, is the largest and most powerful yet built.

A thousand feet in front of me rose the 78-foot rocket-launching tower, its red steel framework silhouetted against the sky. Suspended in the exact center line of the tower, tail a few feet above the ground, nose pointed skyward, was the gleaming silver and black projectile, ready for its flight. It was 21 feet long from its pointed front, down

over its 18-inch shoulder, and along its tapering, streamlined body to the
rear end, four inches in diameter.

Modern Rocket
Powerful as a Steam Locomotive

A far cry from the device that usually comes to mind when the term
"rocket" is used, this was virtually a powerful engine of tanks and pipes
and valves, and high gas pressures. Although weighing with its fuel only
one-five hundredth as much, it generates a horsepower about equal to that
of one of the ponderous steam locomotives that pull fifty or more loaded
freight cars on American railroads.

The last checkings and inspections of the rocket had been completed,
valves had been put in the proper position, the fuels had been poured into
the tanks, and the last workman had hurried away to a safe distance.

I pressed down the first key, and immediately heard the expected sharp
click that told me the prearranged train of events had started. There was
another click, then the whir of small motors, a sound that rose rapidly to
a sharp, screeching whine as the wheels and shafts turned faster and
faster.

A second and a half later there was a terrific explosion with smoke and
flame bursting from the bottom of the rocket. Undoubtedly, those new to
rocket experimentation feared that something had gone wrong. But it was
a planned explosion, and one of a most unusual type. There was no single detonation nor series of detonations; instead a mighty, roaring explosion that went on and on. It meant that a stream of gasoline and a stream
of liquid oxygen, gushing in to the top of the combustion chamber near
the rocket's tail, had met and had been ignited; and that in an almost infin-
itesimal fraction of a second they were flashing into flame.

Gases From Burning Oxygen and Gasoline Rush Out at
Mile-a-Second Speed

The white-hot, expanding gases that resulted had to find their way out of
the chamber, and they were rushing now from the tail of the rocket at a

speed of nearly a mile a second.

Deflected by a hollow concrete elbow that we had dubbed "the bathtub," they went roaring out behind the tower, doing no more harm than to fuse a thin crust onto a patch of New Mexico soil. Such surging gases are extremely powerful, and their power was increasing rapidly as the various parts of the rocket warmed up.

I watched intently through a telescope focused on a group of dials and other indicators near the base of the tower. When the explosion had continued for perhaps two seconds, a yellow light flashed on, indicating that the rocket was lifting itself upward against the metal arms that held it down. But I needed another signal, and soon it came; a flash of red light which told me that now there was an upward thrust on the rocket of 200 pounds more than its weight, and that at last the time has arrived for it to soar into the sky.

"At last" meant in this case exactly seconds, but it has seemed to me, as I stood tensely watching developments, more nearly like ten minutes.

I held back for perhaps a quarter of a second more for good measure, then pressed the vital second key, and the restraining arms snapped out of the way. Immediately the rocket started moving upward along its guiding track. With a long blast of flame roaring downward from its base, and leaving a trail of smoke, it moved slowly for the first length, faster as its tail left the tower, then rapidly gathering speed, it roared practically out of sight, miles into the sky. A considerable time before the greatest height was reached the smoke had ceased and the rocket, still streaking skyward at perhaps 500 miles an hour, was "coasting."

So far the flight was a success, and I watched to see the next chapter unfold. Soon two patches of white could be made out against the blue: a small parachute beneath which dangled the rocket's detachable nose containing the barograph for measuring altitude, and a larger parachute from which hung the remainder of the rocket. These had been released automatically as soon as the rocket reached the top of its flight. Some minutes later both pieces landed in good shape only a few thousand feet from the launching tower, and the test flight was successfully completed.

Flight Marks a Significant Milestone in Rocket Development.

I think I may say that this April flight test marks a significant milestone along a path that I have been traversing for 30 years, a path that has led through fields of study and mathematical analysis, experimentation, invention, laboratory and shop work, ground tests, and air demonstrations. My aim had been to take the crude wood-and-paper rocket which for so long has been merely a toy for Fourth-of-July celebrations, or at best a rough and ready utility for ship masters and life-saving crews, and to transform it into a dependable scientific tool which could lift recording instruments to a greater height into the atmosphere above the earth's surface than has been possible by means of balloons or any other device previously developed.

The measurable success reached does not mean that the problem has been fully solved. It does mean, however, that we are on the right track. It gives us the satisfying assurance that we have solved the problems of how to keep the path of flight approximately straight up and how to recover the projectiles for further use; and that in the field of basic design and use of fuels we have arrived at a point from which we can see more clearly the further developments and simplifications that lie ahead. In other words we have a powerful rocket that works. Now to make it work more effectively.

Launching Must Be Made Independent of Weather

As rocket development has gone forward through the years, I have glimpsed many possibilities for betterment of the fundamental features of the rocket. But I have temporarily put these matters aside, acting on the conviction that if a feature works it should be retained until such time, at least, as all other contributing features have been gotten into workable shape.

Among the refinements that are now most desirable are methods to make the launching of rockets independent of wind and other weather conditions such as rain and snow. The launching arrangements which we are using consist of many devices, each of which has been proved workable, but which taken together now constitute a rather complicated system, and one which is not easily transported nor rapidly set up. If the rockets are to be sent up frequently in various parts of the country to gather weather information quickly, it will be necessary to simplify the launching arrangements so that loading and firing can take place in a very short time. This will make it possible to obtain weather data in a matter of minutes from the entire region of the stratosphere –which is the ceiling for weather. Doubtless this will be on the first practical applications for rockets. Of course the rocket can also pass far into the stratosphere where information can be obtained [that would be] of interest to physicists studying the outer atmosphere, radio engineers, and astronomers.

Our experiments and experience have indicated, to a considerable extent, the lines that future developments should take in order to make rockets a simpler, more dependable, and more effective instrument for covering long distances at very high speeds.

In the matter of pumps, for example, those which we are already using – weighing only 1-1/4 pounds – have given pressures in shop tests sufficient to raise water 1,800 feet, or more than a third of a mile. Such power is extremely important, because the greater the pressures that can be used in the chamber during the combustion of the fuels, the greater the lifting force, resulting in both greater range and greater speed. It is desirable, therefore, to increase our pressures still more.

We can obtain a still further increase in lifting and driving power by altering the design of the chamber in such a way as to burn the fuel to the best possible advantage, and so squeeze from it all the energy possible.

Another line of development is to achieve extreme lightness throughout the construction of the rocket without sacrificing strength. In the case of a rocket whose weight consists almost entirely of fuel, which is the ideal toward which we are striving, the fuel tanks will constitute the largest feature of the structure. A hint of what the future holds for light tank construction is given by recent experiments with fused quartz fibers used in place of piano wire, to wind about thin-walled metal tanks to strengthen them. Already such fibers have been made with tensile strengths of three and a half million pounds to the square inch – ten times as strong as heat-treated steel and only one-third as heavy.

Since the earliest theoretical rocket development, a great many books and articles have been written speculating on the use of the rocket itself and of jet propulsion applied to airplanes, for travel at long range and high speeds, sending mail across oceans, and similar projects. There is sound physical basis for these speculations. In the first place the rocket would travel best above the atmosphere because there is no air resistance, yet this is the region in which an airplane could not be supported. Further, the rocket, as an engine, operates with greater efficiency in a vacuum whereas an ordinary airplane engine and its propeller require air of reasonable density. Moreover rockets – inefficient at low speeds – begin to become really efficient at about the top speed that now seems feasible for airplanes: roughly 500 miles an hour.

Nevertheless, it has seemed to me impractical to go beyond the speculation stage in regard to these matters until we perfect the rocket for meteorological use and gain real engineering information on which to base concrete designs.

Still more impractical would be attempts to construct rockets capable of passing outside the earth's atmosphere, before the problems of rocket design and use have been thoroughly worked out inside the earth's atmosphere.

2

An Interview with Alan Shepard

The First American in Space
Pilot, Mercury Freedom 7
Spacecraft Commander, Apollo 14
Fifth Man to Walk on the Moon
Chief Astronaut Office 1963 - 1969

Americans mourned the loss of their first hero in space, Alan B. Shepard, Jr., who died on July 21, 1998, after a long battle with leukemia. Shepard's spirit lives on, however, in the memories and recollections he shared with the world over the past 40 years. Alan Shepard strove to be the first and the best at everything he did, a goal he usually achieved. One of the best military test pilots in the 1950s, Shepard correspondingly earned his place as one of the first American astronauts. He was part of the "Original Seven" chosen by NASA in 1959, along with M. Scott Carpenter, L. Gordon "Gordo" Cooper, Jr., John H. Glenn, Jr., Virgil I. "Gus" Grissom, Walter M. "Wally" Schirra, Jr., and Donald K. "Deke" Slayton. Of these seven astronauts, Shepard earned the honor of flying the first manned Mercury mission, Freedom 7, on May 5, 1961, and his place in history as the first American in space. To top off his astronaut career, Shepard was the first, and to date the only, person to play golf on the Moon, which marked a colorful aside to his Apollo 14 mission in 1971. Shepard co-authored his book "Moon Shot" with Deke Slayton and so preserved some of his memories for the benefit of future generations. Yet, Shepard did not simply reminisce of his past accomplishments, but he worked toward improving the future by serving as president emeritus of the Astronaut Scholarship Foundation.

A few months before his death in 1998, Shepard granted an interview to Roy Neal.

***

Q: Let's begin...not at the beginning, because there was a beginning before this, but does the date 9 April 1959 mean anything to you?

Shepard: Well, of course, that was one of the happiest days of my life. That was the day on which we all congregated officially as the U.S. first astronaut group. We had been through a selection process obviously previous to that time. But that was the day we first showed up officially as the first astronauts in the United States, back at Langley Field, Virginia.

Q: Your journey to get there took you through test pilot school, took you through combat experience, took you through a bit of everything, didn't it? Why was it that NASA decided to pick test pilots, of all things, to fly the first space mission?

Shepard: Well, I think that it was an immediate realization that we had essentially a new product. It didn't look very much like an airplane, but if you were going to put a pilot in, it was going to have to fly somehow like an airplane, and when you have a strange new machine, then you go to the test pilots. That's what they were trained to do, and that what's they had been doing. Now, of course, NACA (National Advisory Committee for Aeronautics) had some test pilots but they were a little bit older. None of them, I don't think, were in a position where they probably could have competed with the varied background of test flying which most of us had. And, so the decision was made. I don't know. They say that Eisenhower had something to do with the decision because he said, "Well, yes, we need a test pilot." He agreed that NACA, NASA now, didn't have very many test pilots "so let's go to the military and see what they have to offer." Now whether Eisenhower himself was involved in the decision, [I don't know but] apparently the White House was to some degree.

Q: But, the point is, of course, you were named. When first you sized up those teammates of yours, I wonder what your first reactions were to the group.

Shepard: (Laughing). I wondered first of all where these six incompetent guys came from. Seriously, it was not a surprise because several of them had been involved in the preliminary selection process, so I was generally familiar with their backgrounds. Glenn, of course, I had known before; Schirra I had known before because of our Navy connections. So I knew there was a lot of talent there, and I knew that it was going to be a tough fight to win the prize.

Q: It was competitive at that time between the seven of you, wasn't it?

Shepard: Well, it was an interesting situation because, as I say, I was friendly with several of them. And, on the other hand, realizing that I was now competing with these guys, so there was always a sense of caution I suppose, particularly talking about technical things. There was always a sense of maybe a little bit of reservation, not being totally frank with each other, because there was this very strong sense of competition.

Seven guys going for that one job. So on the one hand there was a sense of friendliness and maybe some support, but on the other hand, "Hey, I hope the rest of you guys are happy because I'm going to make the first flight." (Laughter).

Q: You were about to move into a whole new world, or a new non-world, up there in weightless space of which nothing was known. Didn't that frighten you just a little bit? What were your thoughts about moving into a new environment?

Shepard: I suspect my thoughts generally reflected those of the other chaps. But, with me I think it had to be the challenge of being able to control a new vehicle in a new environment. This is a generalization, but it's something which I'd been doing for many, many years as a Navy pilot, as a carrier pilot; and believe me, it's a lot harder to land a jet on an aircraft carrier than it is to land a LM (Lunar Module) on the Moon. That's a piece of cake, that Moon deal! But that was part of my life, the challenge. And here you had, yes, a new environment, but you know, for fighter pilots who fly upside-down a lot of the time, zero-gravity wasn't that big a deal. Since none of us were medics we hadn't thought about the long-term effects of zero-gravity, but the short-term effects of zero-gravity were not the challenge to us. The challenge was to be able to fly an unusual craft and provide good, positive, thinking control of that vehicle.

Q: So unusual a craft that there weren't even any training devices or simulators that could simulate the kind of things you were going to do. You had to make them didn't you?

Shepard: Well, you know that's exactly correct. In the early days we really had what we called "part-task trainers" instead of simulators. Something was built to indicate the control system; something else was built to indicate the radio systems or some of the instruments. And they were all sort of separated, not the great, glorious simulators which we have today.

Q: What was the role of the astronaut in those devices?

Shepard: Well, I think that the role of simulators then, today, and tomorrow has to be: you're dealing with individuals who fly unusual aircraft, who conduct unusual experiments infrequently, because you don't fly in space every day. So there has to be the simulator, which artificially creates problems for you to train against or train with, to learn how to overcome difficulties you may be having with your experiment, difficulties you may be having with the tail of the Shuttle or that sort of thing. So simulators are a very, very important part of space flight and they're also a very important part of commercial aircraft. Unfortunately, some of the companies today, the commuter [aircraft] companies, don't require simulator time, which is surprising to me. I think many of the pilots do it on their own. But simulators really are good because they create a sense of confidence in oneself. You go up and the engine quits and you land safely; or you go up and the rocket goes sideways and you get out, come back home and do it again. So there's a lot of confidence created in the simulation business.

Q: Did you or the other astronauts take an active role in designing the spacecraft?

Shepard: Yes, we did, and we tried to do it as efficiently as we could. In the early days, with only seven, we assigned an individual to work directly with the contractor. And this was all with NASA's blessing, because the NASA engineers were there as well. But primarily from a pilot's point of view, is this handle in the right place? If you have a switch which you have to use to counteract an emergency, is it reachable, is it visible, or do you have to go behind your back somewhere to find the darn thing? Primarily from a pilot's point of view was our interface.

Q: Then finally you wound up being the first man to fly in a Mercury spacecraft. Did you know that was coming or was it a surprise?

Shepard: We had been in training for about 20 months or so, toward the end of 1960, early 1961, when we all intuitively felt that Robert R. "Bob" Gilruth (Director, Space Task Group) had to make a decision as to who was going to make the first flight. And when we received word that Bob wanted to see us at 5:00 in the afternoon one day in our office, we sort of felt that perhaps he had decided. There were seven of us then in one office. We had seven desks around in the hangar at Langley Field. Bob walked in, closed the door, and was very matter-of-fact as he said, "Well, you know we've got to decide who's going to make the first flight, and I don't want to pinpoint publicly at this stage one individual. Within the organization I want everyone to know that we will designate the first flight and the second flight and the backup pilot, but beyond that we won't make any public decisions. "So," he said, "Shepard gets the first flight, Grissom gets the second flight, and Glenn is the backup for both of these two sub-orbital missions. Any questions?" Absolute silence. He said, "Thank you very much. Good luck," turned around, and left the room.

Well, there I am looking at six faces looking at me and feeling, of course, totally elated that I had won the competition. But yet almost immediately afterwards feeling sorry for my buddies, because there they were. I mean, they were trying just as hard as I was and it was a very poignant moment because they all came over, shook my hand, and pretty soon I was the only guy left in the room.

Q: That's a priceless story, Alan. Finally things progressed to the point where you're getting ready for the flight. And if I'm remembering correctly there were some holds dealing with that day on the launch pad. Let's go back to that day, as you remember it. You're getting ready now for MR-3, as it was loosely labeled.

Shepard: Actually the checkout, the countdown had been running very, very well. Of course, Glenn was the backup pilot and he'd been in on the pre-flight stuff. The Redstone checked out well. We had virtually no problems at all and were scheduled for, I believe it was, the second of May. And, I was dressed, just about going out the door when a tremendous rainstorm, thunderstorm came over and obviously they decided to cancel it, which I was pleased they did. It was rescheduled for three days later, and of course, went through the same routine. The weather was good, and I remember driving down to the launching pad in a van which was capable of providing comfort for us with a pressure suit on and any last-minute adjustments in temperature devices and so on that had to be made; they were all equipped to do that. The doctor, Bill Douglas, was in there. We pulled up in front of the launch pad. Of course, it was dark. The liquid oxygen was venting out from the Redstone. Searchlights all over the place. And I remember saying to myself, "Well, I'm not going to see this Redstone again." And you know, pilots love to go out and kick the tires. It was sort of like reaching out and kicking the tires on the Redstone because I stopped and looked at it...looked back and up at this beautiful rocket, and thought, "Well, okay buster, let's go and get the job done." So I sort of stopped and kicked the tires, then went on in and on with the countdown.

There was a time during the countdown when there was a problem with the inverter in the Redstone. Gordon Cooper was the voice communicator in the block house. So he called and said, "This inverter is not working in the Redstone. They're going to pull the gantry back in, and we're going to change inverters. It's probably going to take about an hour, an hour-and-a-half." And I said, "Well, if that's the case then I would like to get out and relieve myself." We had been working with a device to collect urine during the flight that worked pretty well in zero-gravity but it really didn't work very well when you're lying on your back with your feet up in the air like you were on the Redstone. And I thought my bladder was getting a little full and, if I had some time, I'd like to relieve myself. So I said, "Gordo, would you check and see if I can get out and relieve myself quickly?" And Gordo came back, I guess after there were some discussions going on outside, it took about three or four minutes-and finally [a voice] came back and said (in a German accent), "No," he says, "Von Braun (the German rocket scientist responsible for the development of the Redstone rocket) says, 'The astronaut shall stay in the nose cone.'" So I said, "Well, all right, that's fine but I'm going to go to the bathroom." And they said, "Well, you can't do that because you've got wires all over your body and will have short circuits." I said, "Don't you guys have a switch that turns off those wires?" And they said, "Yeah, we've got a switch." So I said, "Please turn the switch off." Well, I relieved myself and, of course, with a cotton undergarment, which I had on, it soaked up immediately in the undergarment and with 100% oxygen flowing through that spacecraft. I was totally dry by the time we launched. But somebody did say something about me being the world's first wetback in space. (Laughter).

Q: At that time the whole game was totally competitive, not alone among the seven astronauts, but you were in a race for space with the Russians. They kind of beat you to the punch, didn't they? I'm thinking of Yuri Gagarin (Soviet cosmonaut who flew in Vostok 1 on April 12, 1961, making him the first human in space) when I say that.

Shepard: That little race between Gagarin and me was really, really close. Obviously, their objectives and their capabilities for orbital flight were greater than ours at that particular point. We eventually caught up and went past them, but as you point out it was the Cold War, there was a competition. We had flown a chimpanzee called Ham in a Redstone-Mercury combination, and everything had worked perfectly except there was a relay which at the end of the powered flight was supposed to eject the escape tower, because it was no longer needed, separate it from the Mercury capsule and eject it. For some reason with Ham's flight, it fired but it did not separate itself. So the chimp was lifted to another 10 or 15 miles in altitude and another 20 or 30 miles in range. There was absolutely nothing else wrong with the mission. So our recommendation, strongly, was, "Okay, let's put Shepard in the next one. Everything worked fine, so if the thing happens again, no big deal. Shepard goes a little higher." Werner said (in a German accent), "No, we want everything absolutely right." So we flew another unmanned mission before Gagarin flew, then his flight, and then mine, so it was really touch-and-go there. If we'd put me in that unmanned mission, we would have actually flown first. But it was tight.

Q: In retrospect it doesn't seem that important, but at the time I guess it was.

Shepard: Oh, very important; absolutely, absolutely.

Q: Then, as time went on you started lobbying for another flight in Mercury, but then Mercury was cut a little short because there was the pressure of something else, wasn't there? Can you discuss those pressures?

Shepard: You are not surprised that I wanted to fly again, are you, Mr. Neal?

Q: Not at all.

Shepard: After Cooper finished his day-and-a-half orbital mission there was another spacecraft ready to go. My thought was to put me up there and just let me stay until something ran out; until the batteries ran down, until the oxygen ran out, or until we lost a control system or something. Just an open-ended kind of a mission. And so I recommended that and they said that they didn't expect to hear anything else from me. But I remember when Cooper and his family and the other astronauts and families were invited to the White House for cocktails with (President John F.) Jack Kennedy, and we stopped at Jim Webb's (NASA Administrator) house first and had a little warm-up there, and I was politicking with Webb and I said, "You know, Mr. Webb, we could put this baby up there in just a matter of a few weeks; it's all ready to go. We have the rockets. Just let me sit up there and see how long it will last, get another record out of it." "Well," he said, "No, I really don't think so. I think we've got to get on with Gemini." And I said, "Well, I'm going to see the President in a little while. Do you mind if I mention it to him?" He said, "No, but you tell him my side of the story, too." So I said, "All right."

So, we got over there and we were all sipping our booze, getting some of our taxpayers' money back drinking at the White House, and I got Kennedy aside and said, "There's a possibility we could make another long-duration Mercury flight, maybe two, maybe three days, and we'd like to do that." He said, "What does Mr. Webb think about it?" I said, "Webb doesn't want to do it." So he said, "Well, I think I'll have to go along with Mr. Webb."

Q: It made you realize who was the power behind the throne.

Shepard: At least I tried.

Q: So instead you started then getting ready to fly in Gemini, another whole new ball-game.

Shepard: Yes. It was very fortunate I was chosen to make the first Gemini mission. Tom Stafford, who is a very bright young guy, was assigned as co-pilot, and we were already into the mission, already training for the mission. We had been in the simulators, as a matter of fact, several different times. I'm not sure whether we'd looked at the hardware in St. Louis or not prior to the problem which I had.

The problem I had was a disease called Ménière's; it is due to elevated fluid pressure in the inner ear. They tell me it happens in people who are Type A, hyper, driven, whatever. Unfortunately, what happens is it causes a lack of balance, dizziness, and in some cases nausea as a result of all this disorientation going on up there in the ear. It fortunately is unilateral, so it was only happening with me on the left side. But it was so obvious that NASA grounded me right away, and they assigned another crew for the first Gemini flight. So there I was, what do I do now? Do I go back to the Navy? Do I stick around with the space program? What do I do? I finally decided that I would stay with NASA and see if there wasn't some way that we could correct this ear problem. Several years went by, there was some medication which alleviated it, but I still couldn't fly solo. Can you imagine the world's greatest test pilot has to have some young guy in the back flying along with you? I mean, talk about embarrassing situations! But, as a matter of fact, it was Tom Stafford who came to me and said he had a friend in Los Angeles who was experimenting correcting this Ménière's problem surgically. And so I said, "Gosh that's great. I'll go out and see him." So he set it up. I went on out there. The fellow said, "Yeah, we do. What we do is we make a little opening there, put a tube in so that it enlarges the chamber that takes that fluid pressure, and in some cases it's worked." And I said, "Well, what if it doesn't work?" And he said, "Well, you won't be any worse off than you are, except you might lose your hearing. But other than that..." So I went out there under an assumed name.

Q: What was the name?

Shepard: It was Poulos, I think. Victor Poulos. The doctor knew and the nurse knew, but nobody else knew. So, Victor Poulos checks in and they run the operation...it's not that traumatic, obviously, because after about a day I was out of there. Of course, it was obvious when you look at the big ball of stuff over my ear when I get back home. But NASA started looking at me. And several months, several months went by, and they finally said, "Yes, all the tests show that you no longer are affected by this Ménière's disease." So there I was, having made the right decision.

Q: Once you went into Gemini all of a sudden there were two of the seven who had been grounded, Deke (grounded for an irregular heartbeat) and Al. What a team. How did it come about that you wound up becoming Chief of the Astronaut Office while Deke, by this time, had assumed quite some power as Head of Astronaut Affairs?

Shepard: Well, as I'd indicated earlier, I decided to fight this Ménière's to stay with NASA. And during the time period when I was grounded, I could become very, very useful in the astronaut training business. I suppose that we really had grown if you'd consider the number of chaps that were involved in the simulators, for example, in the suiting procedures, taking care of the suits and so on, direct supporting the facilities for the astronauts; there were really quite a number of people involved. So they decided to make it a separate division. Deke was the head of that division, and I was given the job specifically of the care and feeding of these astronauts, in charge of their training, helping Deke with crew assignments, that sort of thing.

Q: Was it Deke primarily that got you the job, or was it just the fact that you had all the qualifications? How did that work?

Shepard: Well, I think it was just a matter of saying, "What do we need?" When I became grounded and informed NASA I was going to stay there, then we had two guys that really...either one of us could have done the job. One little difference I think, perhaps that I knew that somehow something was going to happen soon with me. I was either going to get the ear fixed or I was gone. With Deke, I think he was more or less resigned at that stage to the heart murmur business, and the medics kept giving him a bad time about that. So I think it was really that Deke probably was more of a long-term commitment than in my particular case, so I think that's really why it was established. You know we just talked it over with Kraft (Christopher C. Kraft, Jr., Flight Operations Division Director) and Gilruth and they sort of agreed that I was a good selection.

Q: You two had quite a reputation for running a very tight ship.

Shepard: Well, of course, Deke and I were both mad because we were grounded. We'd both been training as astronauts. We knew where every skeleton was in the whole process, and we just wouldn't let those guys get away with anything. We knew what they had to do, we knew how they had to do it, and if they weren't doing it then we would bring them in and tell them about it. Maybe I was a little more forceful than I would have been normally, because of being grounded. I believe they called me The Icy Commander or some "friendly" term like that.

Q: Steely-eyed?

Shepard: We knew where all the skeletons were.

Q: Knowing that, in a very peculiar way from a NASA point of view, perhaps it was for the betterment of the space program that you and Deke were both doing what you were doing at the time you were doing it. Did you ever think of that?

Shepard: I think certainly there was a need for coordination, there was a need for representation at the executive level. Other chaps could have done the job perhaps equally as well or perhaps even better. But it seemed like we turned out some pretty good crews.

Q: I don't think anybody could fault your selection of crews, Alan. All the way through the Gemini Program and finally on into Apollo. And it was during the time of Apollo by which time you finally located through Stafford's ministrations, as you described earlier, a way to treat the Ménière's syndrome in Los Angeles. Suddenly the skies opened again for Alan Shepard. Or did they? You had to get back into the program, didn't you?

Shepard: Well, of course, when NASA finally said I could fly again, I went to Deke and said, "We have not announced publicly the crew assignment for Apollo 13. I have a recommendation to make." I had picked two bright, young guys, one of them a Ph.D., and one of them a heck of a lot smarter than I was, and made up a team to go for an Apollo flight. I said, "I would like to recommend that I get Apollo 13, with Stu Russo as Command Module pilot and Edgar D."Ed" Mitchell as lunar pilot." Deke said, "I don't know. Let's try it out." So we sent it to Washington, and they said, "No, no way." So we said, "Now wait a minute. Shepard's got to be at least as smart as the rest of these guys, maybe a little smarter." And they said, "Well, we know that. But it's a real public relations problem. Here this guy's just gotten un-grounded and all of a sudden, boom!, he gets a premier flight assignment." So then the discussion went on for several days and finally they said, "All right, we'll make a deal. We'll let Shepard have Apollo 14. Give us another crew for Apollo 13," and so that's what happened.

Q: Oh, and did it ever! Suddenly Apollo 13, on its way to the Moon, ran into huge problems. I wonder what you thought when the problem developed, and what did you do during that time period?

Shepard: Well, of course, the immediate thought was, "How do we get these guys back?" Obviously right from the start it was the end of a landing mission; no question about that. But it was interesting to see the entire system being flushed out, being made to come back with any kind of a recommendation. And, of course, Chris Kraft and Gene Kranz were the guys that held everybody together on this thing and said, "Look we've got to find a way to bring these boys back. Failure is not an option." And as you well know, the whole system was vibrating. In any corner of the manufacturing processes, the vendoring processes, NASA's people, everybody was working toward a solution for this problem. As it turned out, there was more than one solution. I mean several different areas of engineering had to be addressed and corrected. And I think it's probably

NASA's finest hour, when you think about it. Certainly from a pilot's point of view, it was just as important an event as stepping on the Moon on Apollo 11.

Q: You had the next flight. Did you approach it with fear and trepidation, or did you approach it with the knowledge that you probably were going to make a pretty good flight out of it, thanks to what had been learned from Apollo 13? Which way was it?

Shepard: Well, ...I know people have expressed the opinion that it might have been a little more dangerous to fly on Apollo 14 than it would have been had there not been Apollo 13. But, recognize that almost a total redesign had to be done; well, not necessary redesign but a total reassessment of the package had to be done, to find out specifically why the thing blew and to fix that, to look for similar situations throughout the service module, but again to reassess the whole scheme of things. You know, in missions like that where you're in basic research there are always decisions along the way; that, well, maybe we should fix this particular piece of equipment because the chances it might fail are one out of 100. On the other hand, it's only a small part of a huge process scheduled to go at a certain time, and if this fails we have the crew to back it up. There are always these little decisions to make, so obviously part of the assessment process of Apollo 13 had to be to go over those decisions again. Now did we have the time to make some corrections of those one in 100 chances of failures? And, of course, several were made in addition to the corrections of the basic problem. So there was a feeling of security, and we were obviously a part of the process.

Q: By that time, too, I had forgotten you had been through the trauma of Apollo 1 and the fire and the redesign that that occurred. Let's go back over that for a moment or two.

Shepard: Yeah, talk about feelings!

Q: Because that must have been a tough one.

Shepard: Well, of course Apollo 1 came as a real shock, no question about it. It came as a shock because it was unexpected, and I'll get into the reasons for it being unexpected a little bit later. But to lose a crew in a ground test, while they're still sitting there on the ground, to lose a crew really woke everybody up. And that was important, because all of us, every single one of us, and Deke and I discussed this, unfortunately after the fact, but we were part of a group that had gone through Mercury, had gone through Gemini, man, we thought, we're leading! We're beating the Russians! We thought nothing could go wrong. And it led to a sense of false security, no question about it. Deke and I remember talking about it. Gus would come back and he'd have a complaint about this. He said, "This is the worst spacecraft I've ever seen." He complained about that. And, of course, he was complaining to engineers as well as to Deke and to me. But Deke and I insidiously became part of the problem because we said, "Okay, Gus, go ahead and make a list of this stuff and we'll see that it's fixed by the time you fly." Not that, "We'll see it's fixed before they stick you back in there for a test where you're using 100% oxygen." You see, there was that sense of security, a sense of complacency that everyone had, including myself and including Deke. I think some people felt that sense of responsibility and neglect, bad decisions, more than others and were personally affected by it more than others. But I don't believe there were more than just a few hardheads that didn't feel in the long run that they were part of the problem.

Q: As it worked out, perhaps because of Apollo 1, Apollo went on to be a hugely successful series of flights.

Shepard: Oh yeah. I don't think there's any question about the fact that the Apollo 1 fire did shape up the whole system, did make people realize that they had been too complacent, that they were over-confident, and it resulted, of course, in a total redesign of many of the parts of the space-craft and, I'm sure, contributed to what was a very highly successful (pro-gram). You know, we're still basic researchers, still putting people on the Moon and you do it six times and you only miss once. I mean that's incredible.

Q: You were really there when the flight to the Moon was born. Wasn't that right about the time following your first, very successful sub-orbital mission? Tell us about it.

Shepard: Well, you know it's an interesting thought, and I've heard it expressed a few times that the decision Jack Kennedy made to go to the

Moon was made after we only had 15 minutes of total space flight time. A lot of people chuckle and say, "Sure!" But the fact of the matter is that, that it's true. And this is how it happened.

We were invited back to Washington after the mission, and I got a nice little medal from the President, and which by the way he dropped. I don't know whether you remember that scene or not, but Jimmy Webb had the thing in a box and it had been loosened from its little clip, and so as the President made his speech and said, "I now present you the medal," and he turned around and Webb leaned forward, and the thing slid out of the box and went to the deck, and Kennedy and I both bent over for it. We almost banged heads. Kennedy made it first...and he said, in his damn Yankee accent,..."Here, Shepard, I give you this medal that comes from the ground up." (Laughter). Jackie (First Lady Jacqueline Kennedy) is sitting there, she's mortified and said, "Jack, pin it on him. Pin it on him!" So he then recovered to the point where he pinned the medal on and everything was fine, and we had a big laugh out of that. But originally Louise and I were supposed to proceed to the Congress after the White House ceremony and then have a reception, and then leave town. But Jack said, "No, I want you to come back to the White House, have a meeting, and let's talk about your flight." So we had the reception at the Hill, drove back, [and] in the Oval Office there were the heads of NASA and the heads of the government. Jack, of course, was there; and Lyndon Johnson was there.

And there's a picture of me sitting on the sofa, Jack is in the rocking chair, and I'm telling him how I was flying the spacecraft, and he's leaning forward listening intently to this thing. We talked about the details of the flight, specifically how man had responded and reacted to being able to work in a space environment. And toward the end of the conversation he said to the NASA people, "What are we doing next? What are our plans?" And they said, "There were a couple of guys over in a corner talking about maybe going to the Moon." He said, "I want a briefing."

Just three weeks after that mission, 15 minutes in space, Kennedy made his announcement: "Folks, we are going to the Moon, and we're going to do it within this decade." After 15 minutes of space time! Now, you don't think he was excited? You don't think he was a space cadet? Absolutely, absolutely! People say, "Well, he made the announcement because he had problems with the Bay of Pigs, his popularity was going down." Not true! When Glenn finished his mission, Glenn, Grissom, and I flew with Jack back from West Palm Beach to Washington for Glenn's ceremony. The four of us sat in his cabin and we talked about what Gus had done, we talked about what John had done, we talked about what I had done. All the way back. People would come in with papers to be signed and he'd say, "Don't worry, we'll get to those when we get back to Washington." The entire flight. I tell you, he was really, really a space cadet. And it's too bad he could not have lived to see his promise.

Q: When he first made that announcement, what was your personal reaction?

Shepard: Oh, we were delighted, [we] were delighted! But there was also a little bit of a gulp in there, because he put a time cap on the deal. I don't think that any of us thought [in 1961] that we would be able to make it within eight years. [We were] delighted but...maybe the President is a little enthusiastic.

Q: We've finally got up to that point where we're into Apollo, and what was your choice, yours and Deke's, what was your best bet as to which would be the first flight to make a manned landing on the Moon?

Shepard: Well I suppose that we felt the schedule as it was laid out, after we rescheduled the Apollo 8 mission, I think that we felt that the missions 9 and 10 adequately demonstrated the Lunar Module's capabilities, that we really deep down inside felt that we could make it. We had a very good possibility of making it on the first try.

Q: And, of course, you did.

Shepard: Of course, we did.

Q: And then along came 14. Because now you had picked your team and you had sweated out Apollo 13, and you were ready to fly. It must have been a big moment when you were ready, waiting for take-off.

Shepard: Well, I think that in retrospect, the obvious advantage here was that Apollo 13 gave us more time to train, no question about it; not that we would not have had enough. But, it gave us a little higher level of comfort with that extra training time. I think obviously the changes to the spacecraft were good ones; not only the changes which related directly to the explosion but others that were made as well. There was a lot of confidence. As I said, I picked a couple of bright guys to go along with me, and there was really a lot of confidence. Eugene A."Gene" Cernan, of course, was my backup. There's a funny story about Cernan.

We were at the point, I think, approximately four or five days away from scheduled lift-off, we were all in quarantine, of course, at the Cape; at that time we had to do the 21 days before, 21 days after routine because of the bug stuff, and Cernan was out early in the morning flying a helicopter, because all the commanders used helicopters to train for the last few hundred feet of a landing. We were having breakfast and we knew Gene was up flying a helicopter, and all of a sudden the door opens and in walks Cernan. He is absolutely covered in soot. He's got scars on his face. We said, "Cernan, what happened?"

He had been flying the helicopter over the river, which was absolutely calm that early in the morning, like a mirror, and he had been distracted by something or other because he was looking at the land instead of the water. He flew that helicopter right into the water, nose over, blades all over the place, tail rotor blades all over the place. Cernan is going down like this. And, of course, being a good Navy-trained pilot, he knew how to cope with being under water. So he got out and he swam to the top and realized he was in fire, so he splashed around and took a big, deep breath and swam a while; and came up, then splashed around some more and swam awhile. He finally got out of the smoke and flames and all that stuff.

Somebody had seen the crash obviously, because the Banana River isn't that big a deal. But he came on the shore, came out and there he was, and just totally bedraggled. So he looks at me as my backup pilot and said, "Okay, Shepard, you win. You get to go." (Laughter).

Q: Alan, you're now on the Moon. You've gotten there on Apollo 14. I wonder what your feelings were the moment you landed.

Shepard: Would you like me to tell you the story about how I got there first?

Q: Oh, yes, of course.

Shepard: Actually, the flight had gone extremely well. We'd had one or two docking problems earlier, a problem with something floating around in the abort switch, which closed as if we were pushing the abort switch closed. All of these were taken care of. Now we're on our way down, flying up on our backs...with the engine pointed that way, slowing down, and getting gradually more steeper and more steeper. We had a ruling that the computer had to be updated by the landing radar; reason being is that while you're on your back obviously you can't see the ground, you can't see the mountains, you can't see the rocks, or anything. So we had a rule that said if the landing radar was not updating the computer by the time you were down at a level of about 13,000 ft, then you have to abort; you have to get out of there. Well, the landing radar wasn't working. They called us up and said, "Your landing radar is not working." We said, "Thank you very much, we're aware of that." And then a little bit further on they said, "You know what the ground rule is about aborting if you're at 13,000 ft." Well, yeah, we knew that. Finally some bright young man over in the (Control Center) said, "Hey your landing radar is working, but it's locked up on infinity. Have them pull the switch, reset it, and see if it works."

So we pulled the circuit breaker, put it back in, and sure enough the landing radar came on. And shortly after that we got cleared to land with a margin of 1,000 feet or so, which was a close thing. As soon as we pitched over there was beautiful Fra Mauro, just the way I had seen it hundreds of times from the scale model. We came on in, made a very, very soft landing. As a matter of fact soft enough so that even though we'd landed in a slight crater, the uphill leg didn't crush like it was supposed to. We had crushable material in the lining. It was a slight ring wing down perfect landing. We shut off the switches and Ed Mitchell turned to me and said, "Alan, what were you going to do if the landing radar had not been working by 13,000 feet?" I looked at him and I said, "Ed, you'll never know."

I would have gone down. I'd come that far. You see, Ed, for example, had not been in the landing simulator at all. It was my job to land. And I'd done hundreds of these things. I knew that if I could see the surface, man, I could get down, maybe not exactly where we were supposed to but

I could get down close to it.

Q: And so you would have made the landing under any circumstances? You'd have broken the mission rules?

Shepard: I would have at least been able to take a visual look. I would have pitched over and taken a visual look before and then made a decision.

Q: Fair enough. Well, we finally have you on the Moon. Mission accomplished. Or was it? Tell me about what you and Ed did on the Moon as you remember it. What were the highlights?

Shepard: Of course the first feeling was one of a tremendous sense of accomplishment, I guess if you will. A tremendous sense of realizing that, "Hey, not too long ago I was grounded. Now I'm on the Moon." There was that sense of self-satisfaction immediately. But then that went away, because we had a lot of work to do. But I'll never forget that moment.

Another moment which I will never forget is after Ed had followed me down and we had set out some of our equipment, taken the emergency samples, we had a few moments to look around, to look up in the black sky, a totally black sky, even though the Sun is shining on the surface it's not reflected, there's no diffusion, no reflection, a totally black sky and seeing another planet: planet Earth. Now planet Earth is only four times as large as the Moon, so you can really still put your thumb and your fore-finger around it at that distance. So it makes it look beautiful; it makes it look lonely; it makes it look fragile. You think to yourself, just imagine that millions of people are living on that planet and don't realize how fragile it is. I think this is a feeling everyone has had and expressed it in one fashion or another. That was an overwhelming feeling in seeing the beauty of the planet on the one hand but the fragility of it on the other.

Q: Being Alan Shepard, of course, shortly after that golden moment you decided to play a little golf.

Shepard: I didn't decide to play a little golf. That is a long story. I will not tell the whole story...so far I'm the only person to have hit a golf ball on the Moon. Probably will be for some time. And being a golfer, I was intrigued before the flight by the fact that a ball with the same club head speed will go six times as far. Its time of flight, I won't say "stay in the air," because there's no air, its time of flight will be at least six times as long. It will not curve, because there's no atmosphere to make it slice. And I thought, "What a neat place to whack a golf ball!" Well, when I went to Bob Gilruth to tell him I wanted to hit a couple of golf balls, (he said) there was absolutely no way. I explained that it was not a regular golf club; it was the handle that we used that we pulled out with a scoop on the end to scoop up samples of dust with. That was already up there to be thrown away. Then we had a club head which I had adapted to snap on this handle and two golf balls, for which I paid: the two golf balls and the club at no expense to the taxpayer. The thing that finally convinced Bob was when I said, "Boss, I'll make a deal with you. If we have screwed up, if we have had equipment failure, anything has gone wrong on the surface where you are embarrassed or we are embarrassed, I will not do it. I will not be so frivolous. I want to wait until the very end of the mission, stand in front of the television camera, whack these golf balls with this makeshift club, fold it up, stick it in my pocket, climb up the ladder, and close the door, and we've gone." So he finally said, "Okay." And that's the way it happened.

Q: Now, some general questions, if we may. Well, I guess we'd better get you back from the Moon. We can't just leave you up there. You played golf; now you closed the hatch and you came back. After that, it wasn't too long thereafter that you finally decided you'd completed your run with NASA. You moved on to other fields.

Shepard: Well, as you recall, of course, the only scheduled missions were the Skylab missions. The crews were already assigned to the joint mission with the Soviets.

Q: Including your friend Deke Slayton. Deke finally got his shot at it, didn't he?

Shepard: Boy, we were so pleased. We were so pleased, bless his heart. Can you imagine having to learn to speak Russian to go into space? I mean, that is above and beyond the call of duty. But he did it. I'm not sure the Russians understood him, but he did it. We were really so pleased and so happy for him.

Q: John Glenn is about to fly again. You and he are pretty close to the same age. I wonder what your thoughts are about John flying.

Shepard: John is a couple of years older than I am; I believe he's seventy-seven. But, I've been saying for years that the taxpayers didn't get their money's worth out of Glenn because he made one flight and immediately went into the Congress. And as a taxpayer, I objected to that. I've been telling John this for years and years. I called him up the other day after the announcement and I said, "John, I'm glad that you're going to give me one more flight for my tax dollars!" (Laughter). I think it's good, quite frankly.

Obviously there are a lot of things about how weightlessness treats individuals, and the person's reaction to weightlessness is both a function of the amount of exercise or lack thereof, their general physical conditioning, and the kind of things that one really needs to know if you're going to be in a long-term mission. The more you find out, the better shape you'll be in. So he's a good data-point. He thinks he's in pretty good shape, and he probably is. But his bones are still more brittle, obviously, and I'm sure that there will be some lessons learned even during that short period of time by looking at his general physical condition, before and after. I think it's a good thing. I think we'll learn something from it.

Q: Looking back on it, what do you think now about the *Life* magazine contract? Good, bad, or indifferent? You don't have to answer that one if you don't choose to.

Shepard: Well, I don't know. With respect to the contract we had with *Life* magazine, I think there's a little ambivalence there. At first it was attractive to us because it provided controlled access to the press. Especially on personal things: on personal relationships within the household; personal feelings of the wives—"how do you feel about your husband going into space;" and that sort of thing. None of us had been involved in any of that sort of publicity or recognition before. And in the early days, it got to be a little bothersome, quite frankly, so I think at the start it appeared to be a way to get around that. And so, it seemed to be welcomed from that point of view. But, then the criticism came about the amount of money involved. So I think all in all, we came out about even. Half the people thought it was a good deal and half the people thought it was a bad deal.

Q: Alan, I've asked an awful lot of questions both from my own point of view and those in Houston. It seems to me it's high time we let you say anything that you'd like, if here's something that we haven't asked that should have been asked. If so, fire for effect.

Shepard: It's been a great part of my life, to be involved in the space program. Even before that, as a Navy test pilot, I had some really exciting, satisfying jobs. But I guess I would have to say that it has been a distinct pleasure to be involved in the space program, specifically in being allowed to make a couple of really recognizable, spectacular, lucky missions.

I think that the thing that has impressed me the most about the whole NASA process is that it has worked so well over the years. When you take a look at a group of civilian engineers and scientists who have to work with contractors who are paid and work for somebody else, who also have to work with the military because you've got the military involved, and that things have really turned out remarkably well. Now there have been some heated discussions between the advantages of manned space flight and unmanned space flight, because there are parts of NASA, as you know, totally dedicated to unmanned space flight. There have been some noted discussions and differences of opinion between the engineers on space flight who would like to automate everything and leave the pilots out of there. But you know in the final analysis, I can't remember any of these decisions that were made with an absolute hard-over judgment.

Always have been, and still are, discussions going on to get the best possible answer. When you take a look at the NASA organization, 1958/1959, nobody would have thought what it has done over the years. Nobody would have thought that the computers which took us to the Moon and back were the forerunners of today's chips and today's technology because of the money and the effort that NASA spent back in the

'60s. Sure we would have computers; no question about it. But we wouldn't have advanced, we wouldn't be at the position we are today without that tremendous impetus that NASA had in making the computers. Satellites, the incredible spate of information flowing back and forth from satellites all springing from the NASA organization. It's remarkable what the organization has done, and is still doing. It's just a great process.

About the Interviewer
Roy Neal was an NBC-TV correspondent who covered the space program in the 1960s. He is the author of "Ace in the Hole", a history of the Minuteman missile.

An Interview with
Pete Conrad Jr.

Pilot, Gemini V
Commander, Gemini XI
Commander, Apollo 12
Commander, Skylab II
Third Man to Walk on the Moon
One of the Great Jokesters

Just off the main road leading into the Johnson Space Center in Houston, a grove of young trees grows alongside a winding brick path. This past winter the trees shone at night with the white lights typical of the holiday season. All the trees except one, that is. Yellows, blues, greens, and reds adorned a solitary oak near one edge of the copse, causing it to stand out among all the others not for its stature but for its colorfulness. Like its monochromatic brethren, this tree has a small plaque at its base commemorating a fallen member of Johnson Space Center family. It is astronaut Pete Conrad's name found beneath the only multicolored lights in Memorial Grove. This tribute, at the suggestion of Apollo 12 crewmate Al Bean, continues Conrad's life-long maxim "to be the most colorful".

Conrad first made a name for himself as a Navy test pilot, then joined the second astronaut class after barely missing inclusion in the Mercury 7. He went on to fly Gemini V and command Gemini XI, Apollo 12, and the first manned Skylab mission. While Conrad's service record serves as testament to his skill as an astronaut, it was his sense of humor and easy-going command style that set him apart in the minds of his colleagues. Even after leaving NASA, he remained active in space activities, flying, and racing. Pete Conrad died July 8, 1999, from injuries sustained in a motorcycle accident in Ojai, California. These excerpts are from an interview with Charles "Pete" Conrad, Jr. at his Universal Space Lines office on March 21, 1997.

***

Q: What were your expectations of the manned spaceflight program after the completion of the Mercury Program and before President (John F.) Kennedy made the commitment for our nation to go to the Moon?

Conrad: Everybody thought that Project Mercury was just another experimental program. Wally (Walter) Schirra (Jr.) and Jim (James) Lovell (Jr.) were in my Test Pilot School class, and when we all met in Washington for the interviews for Project Mercury, the subject of the conversation was more of what it would do to our career track. It would take three or four years out of our normal Navy career to go ride around in Mercury. There was no plan at the time to follow-on after Mercury. All through 1960 there was no public talk about doing anything after Mercury. It wasn't until 1961 when Kennedy made the announcement that it became obvious something else was going to happen.

Q: "The Other Nine" had the most impact on the Gemini and Apollo Programs. This group had flown on every Gemini mission and 11 of 16 of the missions from Apollo through Apollo-Soyuz Test Project (ASTP). Tell us what made this group a head above the rest in being assigned to the high profile missions.

Conrad: It was just the fact that the Mercury guys were older and "The Other Nine" were specifically brought in for the longer programs. It became very obvious that this would be a career from then on. Deke (Donald K.) Slayton was no longer able to fly. There is no debate that

(M.) Scott Carpenter had screwed up and Dr. (Christopher) Kraft (Jr.) did-n't want him to fly any more. Wally Schirra flew in Apollo. Gus (Virgil) Grissom got killed in the Apollo 1 fire. John Glenn (Jr.), who was the oldest of "The Original Seven," elected to leave because he didn't want to fly any more missions. Al (Alan) Shepard (Jr.) came down with the ear problem at the beginning of the Gemini Program or he would have been flying all the way through it. Of course, he came back up on flight status and went ahead to fly on Apollo 14. I think that most of the guys – except for John Glenn who had decided to leave and the ones that got killed or couldn't fly – got as far as they wanted to go and quit.

Q: Astronauts are some of the most competitive people on and off this planet. How did your group perceive "The Original Seven" and vice versa?

Conrad: They weren't the happiest guys in the whole world when we came along. It was not so much from the point of view of flying as it was from sharing the perceived goodies that came to them during the course of the program. It became obvious to all of us after we had agreed that we would share and share alike. It was only a year after the nine of us came into the Astronaut Office – bringing the total to 16 – before the arrival of the third group of 14 astronauts and shortly thereafter the fourth group of six scientist-astronauts. The manning of the Astronaut Office quickly went from seven to 16, 16 to 30, and then from 30 to 36. What little of the perceived goodies that existed diminished pretty fast as the Astronaut Corps grew. The third group probably had the same feeling that we did when they arrived. They had felt that the other 16 of us did-n't want to see them around because of the diminishing share of the per-ceived goodies. Obviously, we were all competitive on wanting to fly a mission.

Q: By day five of the Gemini V mission, the crew surpassed Soviet Cosmonaut (Valery F.) Bykovski's endurance record. This was the first time that the U.S. surpassed the Soviet Union in a space record. Describe the crew's attitude and morale after reaching this milestone on the mis-sion. Was the crew too busy to enjoy winning this battle in the space race?

Conrad: That was part of the problem, we weren't busy enough. It made the mission awfully long. We were aware of that target but I don't remember if it meant a heck of a lot to us at the time. We weren't that happy with the mission. By that time we had lost thrusters – drifting along in the flight – and we were flying a medical experiment. Most of the experiments had gone by the wayside because of the power down mode. It made the mission go extremely long.

Q: One of the Department of Defense experiments on the mission was to discern man-made objects and structures on the Earth based on visual observation. This was significant to the U.S. Air Force Manned Orbiting Laboratory (MOL) Program. What were some of the significant observations that were made by the crew?

Conrad: We had two sites. One was in Texas and the other in Australia. It was a visual acuity experiment. They would end up being miserable failures as far as I could tell. We either couldn't find them or the weather was bad. We were supposed to describe in detail the patterns that were laid out on the sites. They had white sheets laid out for us to identify. We had spent a lot of time flying over the Texas site in a C-130 (cargo plane). My recollection was that I really couldn't find much of anything. However, I do remember that there were many things that we could see. That fact verified that a crew could see things on Earth pretty well as long as the lighting conditions were right. For instance, we did find the location of ships at sea by seeing ship wakes that trailed miles and miles from the actual ship. We followed up to where the ship wakes converged and sure enough there was a little ship. We probably wouldn't have found the ship had the wake not pointed our eyes to it because of its small size. But I was impressed with how well we could see from 140 nautical miles (high) in orbit. I remember seeing red roofs in China. We could pick out interstates and large clusters of buildings. We could figure out what cities we were looking at during the night just by the lighting patterns. The only reason for the visual acuity experiment was because Gordo (Gemini V Commander L. Gordon Cooper) had said on his Mercury flight that he had seen all kinds of things on Earth. We only got as far north as southern California, southern Texas, and along the gulf that spanned over to Florida. The Saltan Sea always stood out well because it was very clear out there in the desert. We picked up the Panama Canal from orbit. I think that the visual acuity experiment, per se, was a flop.

Q: You were a big supporter of the Large Earth Orbit (LEO) Program which proposed the use of an Agena stage docked with a modified Gemini spacecraft to fly around the Moon.

Conrad: Sure. That was going to be the flight that I was going to fly. I was all for it. The way it came down was that Mr. (NASA Administrator James) Webb made a very good decision. He said, "Look. Gemini is to support Apollo. You've got to accomplish those things with the Gemini vehicle. Let's not divert our efforts over to do something that's more or less spectacular as opposed to proceeding with the program." I didn't disagree with that. I obviously wanted to fly the flight. Gemini XI would have been the test flight. Gemini XII would have made the LEO. Gemini XI would have simulated the reentry velocities from a LEO mission coming back from the Moon.

Q: It would have been interesting to see if we would have used the LEO Program if the U.S. had significantly trailed behind the Soviet Union in the space race.

Conrad: The things that astronauts did not have access information to, were the black programs that were telling NASA management – the Administrator down to the head of the Manned Spacecraft level – where the Russians were in the space race. Any decisions or recommendations that we made in the Astronaut Office were based on our own desires to do things and move the program forward. It was very obvious from Mr. Webb's decision that he didn't think the Russians were going to beat us. That was more of the reason for cranking up LEO than anything else just to get really out in front of the Russians.

Q: The crew of Gemini V was known to have a solid sense of humor. Share with us some of the humorous moments during training and on the mission.

Conrad: Yeah. We had a sense of humor. The one that I remember was singing during our UHF (ultra high frequency) tests on the radio. I remember singing to the Bermuda tracking station, "Over the station, over the blue. Gemini V singing to you!" I'm sure that we had a lot of laughs but it was so long ago to recall anything specific.

Q: Gemini XI's first EVA had required that a tether be hooked up from the Gemini spacecraft to the Agena by [Pilot] Dick Gordon (Jr.). Gordon became so physically fatigued during the EVA that it was cut short. Were you concerned that the state of the EVA was reaching to a dangerous level?

Conrad: That was probably the only time that I really got concerned on any of the flights. He was obviously in trouble and I didn't want to make it worse. I figured that with my encouragement he had to make up his mind that it was time to knock off and get back in the spacecraft. We made up our minds at the right time. He made the right decision and got back in the spacecraft. We got the hatch closed and everything was fine. That experience made it very obvious that the zero-g airplane had led us down a primrose path.

I had heard some comments from Col. ("Buzz") Aldrin (Jr.) that suggested we had screwed up the EVA but I don't quite look at it that way. My theory in conducting any of the missions that I ever did was that there was never enough time to get things done. That's why I always pressed to stay on or ahead of the flight plan. He felt that we had screwed up on the checklist by being ahead on the timeline, which then led to the difficulties of attaching the sun visor to Dick's helmet.

Q: The crew was excited about being boosted into a higher orbit by Agena's PPS. You had mentioned that it was the biggest thrill in your life. Share with us some of your thoughts on that event.

Conrad: That was going to be very exciting to go higher than anybody. Gemini X – the flight before us – got to 450 miles and we were going to 850 miles, which was the maximum altitude that Agena had in it. We had a lot of problems with it. It was Bill Anders who was on our backup crew that had worked all of that out because he had his masters degree in nuclear engineering. We had some experiments where the scientists were saying, "We're going to screw it all up by going through the radiation belt." It was Bill who had worked out all of the details on burning Agena under the belt. Going up over Australia was picked because that had the lowest radiation level. It was a real thrill. The only unfortunate thing about it was the way it worked out on the flight plan.

Our most spectacular photograph was over India as opposed to over Australia – our highest point in orbit. We couldn't get a good photograph while flying over Australia because we were just about at the terminator and going on into the night. It was a great sight. I also remember saying "Eureka!" Houston, the world is really round." That's when I got all kinds of mail from The Flat Earth Society people in England telling me that I was full of crap.

Q: Compare the flying characteristics and maneuverability of the Gemini spacecraft to the Apollo CSM (Command and Service Module).

Conrad: Well, most of the maneuvering that we did in Apollo was in an automatic mode. There was very little hand-flying. What hand-flying that was done on the CSM was done by Dick. I did all of the hand-flying on the Gemini, although I did let Dick fly it around the Agena before we docked. The Gemini was quite like an airplane. We spent a lot of time learning how to fly it in the pulse mode where we were really in control. We became very proficient at that. I don't remember doing things like that with the Apollo spacecraft because we didn't have to. We did the hand-flying in the LM (Lunar Module) for landing or in the case of a backup mode for docking with the CSM. Normally, the CSM was prime and the LM would come up to the rendezvous before the maneuvering would be turned over to the Command Module Pilot.

Q: In the early days of the Apollo Program, the backup crew for the proposed Apollo 3 mission – an E Mission using a Saturn V for a Lunar Module test – was yourself, Gordon and C.C. Williams. That flight was later redesignated as Apollo 9 but the Lunar Module for the prime crew (Frank) Borman, (Michael) Collins and Anders – was not ready. This prime crew was reassigned to the Apollo 8 mission while your backup crew remained on the Apollo 9 mission. Your crew now served as a backup to McDivitt's crew – McDivitt, (David R.) Scott, and (Rusty) Schweickart – to fly the first LM Earth orbital test flight while Borman's crew eventually flew a C Prime Mission (lunar orbital flight). If your backup crew would have followed Borman's crew to the Apollo 8 mission, you would have been in line to command the first lunar landing. Why wasn't the switch made from Apollo 9 to Apollo 8 with your backup crew?

Conrad: The backup crew for Apollo 8 (Neil Armstrong's crew) and the prime crew for Apollo 8 (Borman's crew) was nowhere as near as schooled on the LM as the prime crew for Apollo 9 (McDivitt's crew) and the backup crew for Apollo 9 (Conrad's crew). We were the first LM trainees even though we were the backup crew on Apollo 9. The initial Apollo 8 prime crew (McDivitt's crew) and my backup crew had the most LM experience. By the time the prime crews were switched, C.C. Williams had been killed and Alan Bean had taken his place. You are absolutely right about the fact that we were not switched over to the Apollo 8 mission. We would have flown Apollo 11 if the flight assignments had gone the way it was originally planned.

Another problem was that McDivitt's LM was late on the production line. They took our CSM while we kept our LM (LM-3) which was the first LM to fly. We picked up their CSM – there was a CSM swap, too. The final crew assignments went along with Deke's philosophy. He always said, "Any Apollo crew was capable of flying any mission." We weren't going to know until shutdown whether or not we would get to fly the lunar landing mission. Everything had to go right. I know because I was sitting in Mission Control behind the Capcom (Capsule Communicator) position when Neil landed. We still didn't know if they were going to make it when the computer glitches popped up. We didn't know until they had finally touched down. I don't have any great shakes about not having made the landing. In hindsight, I would have loved to make the first landing on the Moon but I wouldn't have swapped the flight of Apollo 12 with any other.

Q: Apollo 12 was a more challenging flight for a pilot because of the requirement for a pinpoint landing.

Conrad: It was a result of the fact that they didn't land where they supposed to. Apollo 11 was four miles off target. They changed our landing site and our mission as a result. Up until that time, we were targeted to land in another area that had nothing in it – just geology. They revised our mission as a result of Neil and Buzz not landing where they were supposed to. They gave us a new landing site. That was when they decided that we would go to the Surveyor (lunar vehicle). Nobody talked about doing that up until that time. I believe it was Gen. (Samuel) Phillips that brought the idea up back in Washington. He was running the Apollo

Program at that point in time. That made Apollo 12 a whole different mission.

Apollo 12 was the most interesting mission because we did go to Surveyor. We did the first ALSEP (Apollo Lunar Surface Experiment Package) deployment. We did the day's first useful work in geology. I think that we got a more rounded flight than anybody else did. I didn't need to go roaring over a lot of territory in a car to make a more interesting flight.

Q: Compare your experiences with flying the Lunar Landing Training Vehicle (LLTV) in late 1967 and early 1968 to the actual flying characteristics of your LM Yankee Clipper.

Conrad: No doubt about it, the LLTV was so bad. We had wiped out three of the four vehicles. I remember telling Dr. Gilruth (Manned Spacecraft Center Director) about how bad it was flying the LLTV. The first question that Dr. Gilruth would always ask anybody – Neil, myself, and the guys after me – was, "Do you need to fly the LLTV to land on the Moon?" Everybody would say, "You got it! If you could fly that thing then the landing on the Moon was a piece of cake." That obviously was the hairiest part of the whole program. Most people don't realize that.

Q: Apollo 12's launch occurred during thunderstorm activity whereby the booster was struck twice by lightning. This resulted in the platform and fuel cells going offline in the spacecraft. Was there a point during the launch and in-flight phase that the mission would have to be aborted?

Conrad: Actually, we generated the lightning as opposed to being struck by lightning. I probably was in the best position to assess what happened. I think that it's on the voice tape. I thought that we were hit by lightning. Once we got past 60-70 seconds in the flight, the ground was able to get the signal conditioning equipment (SCE) back online. They could now get some telemetry to tell us that we didn't have a short. I made the instantaneous decision when all that happened not to abort. I just got firmer in that decision. Because all three fuel cells had fallen off the line, I thought, "I'd sure hate to wind up in orbit in a dead spacecraft." That eventually resolved itself down to where I thought, "I'm sure we're going to wind up safe but I really don't think we're going to get to go to the

Moon." We had spent a long and anxious time in orbit straightening things out and working with the ground to verify that things were all right. I was really surprised that they let us go.

Q: Do you think the decision to launch in that kind of weather activity was prudent?

Conrad: There was no thunderstorms or lightning activity in the area. In fact, the final weather report was provided by Air Force One that had just landed. It reported that the top layers were at 20,000 feet with steady rain. There were no thunderstorm cells. My concern was that our forward firing thrusters on the Service Module would be full of water. I even asked if they would blow out sufficiently and whether the water would freeze at the bottom of the forward firing thrusters. This could cause the assembly to blow up if an operation was attempted. I was more concerned about the rain than the lightning. I don't think that anybody had the thought of being struck by lightning. The conditions were such that we generated electricity. We acted like our own static wick. The ionized gases on the engines still gave a nice path all the way to the ground.

Q: Do you think that the presence of President Nixon being there had added pressure to go ahead for a launch?

Conrad: No. A scrub on the Moon was a 28-day launch scrub. If the flight didn't make the launch window that day, it couldn't go for a launch the next day. The mission would be scrubbed for 28 days or a full lunar cycle. It wasn't like we had a window of opportunity to launch every day. We did not.

Q: What were some of your thoughts and observations during the critical descent and approach to the lunar surface?

Conrad: We couldn't see anything until we got into the second phase of the descent program, which didn't occur until 7,500 feet. When the LM pitched up at 7,500 feet, I didn't have the foggiest idea where I was. There were 10,000 craters out there. Al (Bean) was calling off the negative angle down through the optical sight which was computer generated. It showed where the vehicle was going to touch down on the lunar surface. The Snowman Crater pattern popped up as soon as I looked at the right angle. I figured that we had it pretty well made.

Q: I believe that you intentionally distanced your point of landing to the Surveyor.

Conrad: The way the (guidance) program was designed it put the target at the center of the crater. Nobody ever landed in the third phase, which was the auto-land phase. It was known as P61, P62, and P63. If we had flown P62 all the way to where it transitioned to P63, we would have stopped over the site at 100 feet. P63 would just allow descent to the last 100 feet. We targeted that as the center of the crater so that we had the most accuracy. We knew that the Surveyor was on the left-rear side of the crate — about the 7-8 o'clock position and halfway down the crater wall – which was where it was supposed to be. We were going to land in the 4 o'clock position so I could see it out of the left window. We would be looking out from the LM's shadow since there would be a little sun angle against our backs. But I never saw it on the way in. It was just natural for me to fly to the right and around the crater because that allowed me the best visual. I would have never gone the other way because I would be in the blind. It was just natural for me to start around the right side of the crater and land in the first suitable spot.

Q: What were the most memorable sights and experiences of your two EVA's on the Ocean of Storms?

Conrad: I don't think that anybody realizes what it means to say that the Moon is one-quarter the size of the Earth until one stands on it. It is small. My first impression was, "I can't believe this. This thing is curving away from me. I can see it going over the hill." We were out walking around, a distance not too far away from the LM, when we realized that we couldn't see it anymore. The LM was 23 feet tall! My reaction was, "Shoot! It's gone." It was later in the mission when Bean said that he didn't know where he was and just followed me. We were out there on the surface with no sight of the LM – no nothing. All we had was our foot tracks behind our direction of travel.

Q: You had an enlarged ballcap made that you could wear over the EMU (Extravehicular Mobility Unit spacesuit) during your lunar traverses. What happened to that?

Conrad: Yeah. They didn't load it on the flight. I don't believe that they did. If they had loaded it, I would have worn it. I had them make it up but I don't know why they didn't load it onboard. Somebody got chicken.

Q: You tried to convince both of your crewmates to follow you over to the Skylab Program.

Conrad: Yes. Al took my advice and transferred over to Skylab. I tried to convince Dick to transfer if he ever wanted to fly again. But Dick wanted to go the last 60 miles. He wanted to take his chances on the prospects of commanding a lunar landing mission (Apollo 18) and walking on the Moon. I didn't think that he would make it. And he didn't.

Q: In retrospect, how do you feel about the school of thought supported by some in NASA management of canceling the Apollo Program after the second lunar landing in order to divert funding to the Apollo Applications Program and (because of) the probability of losing a crew?

Conrad: That must have been way before we landed because AAP up until 1967 had included a Space Station and the plans to go to Mars. It was President (Lyndon) Johnson that shot that all down. By the time that I got to go to the Moon, I knew that the program was in deep yogurt and that I would never go back again. I knew that I had to go over to Skylab if I wanted to fly again. That was it. It wasn't NASA management that decided to quit the Apollo Program but rather the President of the United States. He made the biggest mistake that anybody ever made in the history of mankind.

Nixon reaffirmed that. They gave the opportunity to go to Mars. He shut everything down again and said to do the Shuttle. Get the first three pages of the first volume of the Challenger Accident Report. The whole history of the Shuttle is right in there. The original Shuttle was a two-stage vehicle. The explanation and the reason why it is in front of the Challenger Accident Report was to set the stage for why it had an expendable tank with solids and no fuel in the Orbiter.

Q: After the (first Skylab) crew made the visual assessment of Skylab, what did you think the chances were of saving the Space Station?

Conrad: I was disappointed that we didn't get the parasol out as soon as we got up there with Joe (Kerwin) and Paul (Weitz) hanging out of the CSM. We found the exact cause of the jammed boom since we could clearly see it. We did the best we could by sending television images back to Earth and gave our verbal descriptions of the situation to the ground. I knew that we were going to work on it sooner or later. It just got very exciting because every day we weren't sure on whether we would get to stay or not. We obviously needed that wing. When we got it out, I'd figured that we had made it in pretty good shape.

Q: Couldn't they have channeled some of the power from the CSM to operate the ATM in a degraded mission?

Conrad: They were sending us up with the idea that we would get the wing successfully out. There was no way to put together an improvised power source for the crew to take up in the next few days. It was not a severe loss for the first mission since there was another source of power available. Earlier in the program, we decided to use the CSM power system until the fuel cells ran out of hydrogen and oxygen. We loaded up the tanks as full as possible so that it would give us power from the CSM. During which time, we felt we could conduct a curtailed but respectable experimental program. Though we were quite concerned about what would happen on the later missions, each was planned to last for 56 days for the most part in which the ATM arrays would be the only source of power.

Q: Why did you leave the Astronaut Office? What projects are you involved in?

Conrad: There was nothing let to fly. I had figured that it would be seven years before the Shuttle would fly. I was 43 years old at the time of the Skylab mission. I would have to wait till I was 50 years old to fly again. I was wrong by one year. It was eight years before the Shuttle flew. I would have been 51 years old. I was offered the first ride and I turned it down. I didn't care for the Shuttle. I didn't care for the way management was organizing things back then. Deke had recommended me to take over the Flight Crew Operations Division and they chose to give it to a non-astronaut. They had offered me the Shuttle ALT (Approach and Landing Test) Program. That was all offered to me before

I decided to leave NASA. When I decided to leave, they had offered me the whole program if I'd stay. I told them that it was too late.

Q: You are now Chairman of Universal Space Lines. What is the company working on?

Conrad: What we are going to do is be the first operators of the commercial reusable launch vehicles. I don't want to build it. I just want to buy it. If I have my way, I'll also go fly it so that I can go back into space.

About the Interviewer
Donald Pealer earned a B.S. in Aerospace Engineering at Boston University in 1988 before becoming a Naval aviator

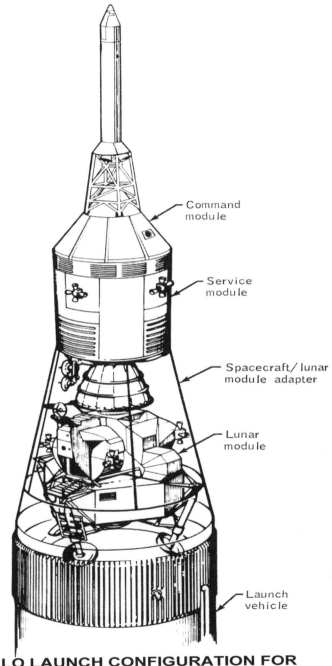

Command
module

Service
module

Spacecraft/lunar
module adapter

Lunar
module

Launch
vehicle

**APOLLO LAUNCH CONFIGURATION FOR
LUNAR LANDING MISSION**

An Interview with
Charles Duke Jr.

Astronaut Support Crew, Apollo 10
CAPCOM, Apollo 11
Backup Lunar Module Pilot, Apollo 13
Lunar Module Pilot, Apollo 16
Backup Lunar Module Pilot, Apollo 17
"Footprints on the Moon – I Can't Believe It!"

Charles M. Duke, Jr. was the pilot aboard the Apollo 16 lunar module "Orion" when it landed on the Cayley Plains in the Descartes region of the Moon in April 1972. This was to be his only spaceflight.

Duke, the tenth human to set foot on the Moon, spent some twenty hours walking and driving on the lunar surface along with his commander, John Young, during three lunar surface EVA's. He also performed a stand-up EVA (SEVA) during the return to Earth phase of the mission when command module pilot Ken Mattingly collected film canisters from the spacecraft's scientific instrument module bay (SIMBAY).

Duke, born in North Carolina on October 3, 1935 became an astronaut in April 1966 as part of NASA's Group 5 selection. He performed support duties on Apollo 10 and was the back-up lunar module pilot on Apollo 13 and Apollo 17. Duke was also CAPCOM on the white team during the first lunar landing mission, Apollo 11, in July 1969. It was Duke who made the reply to Neil Armstrong's "The Eagle has landed," by saying: "Roger, Tranquility, we copy you on the ground. You've got a bunch of guys about to turn blue We're breathing again. Thanks a lot."

Duke left NASA in January 1976 with just over eleven days of space-flight experience, three of which had taken place on the moon.

Randy Attwood and Keith T. Wilson both spoke with retired Brigadier General Charles M. Duke, Jr. (during separate occasions) about his experiences as an astronaut. Both interviews have been combined below. Wilson's questions with corresponding answers are indicated by "Quest*"; Attwood's by "Quest."

***

Q: Take us back, if you would, to getting ready for your flight. What did an astronaut have to do to get ready to fly to the Moon?

Duke: In the Apollo program, it depended on your job. If you were a Command Module Pilot, which meant you stayed in orbit around the Moon while the other two crewman landed, your training was slightly different. You didn't have to worry about the lunar surface activities, the rover, the spacesuit work, things like that. That's what we did. We had to learn about geology, the science of the Moon, the geology of our landing site as best we could determine from the photogeology. We had to learn all about the experiments we had to do on the Moon – and then we had to learn how to fly the spacecraft which was our primary emphasis in training. The Command Module pilot did similar things during his training sequence. He learned how to fly the spacecraft and how to conduct the experiments he was going to do from orbit and then he did a lot of work on the extra-vehicular activity that we had on the way home.

Q: Let's go back to the day you started your trip on April 16, 1972. I can't imagine sleeping the night before going to the Moon.

Duke: It was a busy time during our training, the final frenzied preparations. We got to bed fairly late as I remember. I don't recall any real
problems then. The first night in space I had a very difficult time getting
to sleep, but the night before I had no trouble at all sleeping. I bounced
out of bed the next morning ready to go.

Q: Tell me about launch day preparations. You first had breakfast and
then put on your space suit?

Duke: The first thing the morning of the launch, we had a physical.
After we woke up, you know, we washed our face and took care of those
kinds of things. We then had a physical which was very brief. Just make
sure we could walk and talk primarily, to make sure we didn't have a temperature or anything like that, any fever. Then we went to breakfast and
there was a whole pack of people there to help us get off and encourage
us on our way. Then we went to the suit room to get suited up in our
spacesuits. We spent an hour there before we went out to the launch pad.
We finally hit the pad about two hours before liftoff, as I recall.

Q: Tell me about what it is like to be launched in one of those incredible
Saturn V rockets.

Duke: The Saturn was an immense machine. It was 360 feet tall and 33
feet in diameter and weighed around six and a half million pounds when
it was full of fuel. We took the elevator up the side of the rocket in the
launch tower, climbed in and were strapped in. About an hour and a half
before liftoff we were secure inside. They closed the hatch and gave us
a pressure check so we just lay there and waited for liftoff, hoping the
countdown would continue and that the launch would go on schedule.
It's really not too comfortable in the spacesuit on your back in one-g, so
we were eager to get it off on schedule. We did launch right on schedule.
You're really tense. As the seconds tick off towards liftoff and at ignition
things get very, very tense and very quiet inside. And then at ignition at
eight seconds before liftoff they lit the five engines of the Saturn and we
began to hear a muffled roar and the vehicle started to vibrate. The countdown continued as the engines came up to full power. At T-0 if all was
going well they gave us a launch commit and as they released, we slowly began to move upward like an elevator going up in a large building. It
was vibrating very hard from side to side. That was the attention getter

for me – the vibration in the vehicle.

Q: As the Saturn V accelerated, the g's began to build up. Was it uncomfortable?

Duke: You really didn't notice it. Your attention was on the instruments and the dynamics of the flight. Four and a half g's didn't seem much different from one g for me. The g level was a slow buildup and it was only about four g's for a few seconds. Then the engines shut down and you felt you were in a train wreck as the acceleration went from 4+ g's to zero in a few microseconds.

Q: Describe the early part of the flight in Earth orbit.

Duke: We were in orbit for one and a half revolutions. We went over Kennedy [Space Center] again and then the second time over Australia we ignited our third stage engine to accelerate us out of Earth orbit on the way to the Moon.

Q: Did your busy schedule allow any quick glances out the window?

Duke: I was too busy and took just a couple of glances – once right after we got into orbit and could see the Atlantic Ocean and it was just beautiful – the deep blue of the water and the blackness of space all framed in one window of the spacecraft. The next time we were over Houston, an hour later, I could roll around in zero gravity and look out the hatch window and there was Houston, Galveston Bay, all of the gulf coast of Texas framed in the window – it was spectacular.

Right after we started out we had to retrieve the lunar module from its launch position beneath the command module. As we turned the spacecraft around to put it into position for the docking maneuver, we noticed particles floating away from the lunar module. It looked like a spray of crystallized fluid to us. We were very concerned it was coming from the area of the ascent engine tank. Our first impression was 'there goes the mission –it's leaking gas'. We reported this to Houston and continued the docking. The flight plan was changed and soon after we entered the lunar module and powered up for a short time to check to see if it was fuel leaking out. It was not. Actually it was the thermal paint covering which had been applied incorrectly and was peeling off.

Q: Your two spacecraft were named Casper and Orion. Describe how these names were chosen.

Duke: Ken Mattingly chose Casper—why, I don't know. We selected Orion which is a bright winter constellation. We had to select names because, during the rendezvous phase of the mission or when we were separated in the two spacecraft, Houston had to call us something. They could have called us Apollo A and Apollo B but that wasn't very dramatic or romantic so we selected names that would have the connotation of space so we picked Orion.

Q: Tell me about the three-day trip to the Moon. What did you do?

Duke: There was a lot of activity that we had to devote ourselves to. Housekeeping chores took up a major part of the time. We had to learn to live together. We had to unstow our food, clothing, and waste management system and put everything back in the proper place. We had some sleeping bags we had to unstow in the evening. The first day or so was just getting adjusted to one another and to the environment of the Command Module. We had a few experiments that first day and a few mid-course corrections to make sure we were right on track. Then we started into some experiments. We had an electro-phoresis experiment; we observed some cosmic particles that were bombarding us during the Light Flash experiment. We had some photographic work to do of deep space. Other than that it was just a matter of maintaining our timeline and waiting with great expectation the day we entered lunar orbit.

Q: How was it for you personally adapting to zero gravity?

Duke: During the first few hours I thought this was going to be a disaster for me. I had the distinct impression that I was going to be seasick. I had gotten seasick while I was at the Naval Academy every time I went out on a ship, so I said eleven days of being seasick is going to be terrible. Fortunately, a couple of hours later, I noticed that the nauseous feeling had gone away and I was in great shape bouncing off the walls with John and Ken. They didn't have any trouble at all. To sleep, I had to take a light sleeping pill to sort of put my mind in idle so that I could calm down and go to sleep. My mind was racing 90 miles a minute, it seemed like, the whole time we were airborne; thinking about the next day's activities, thinking about the beauty of space, looking at the sights, worrying about the spacecraft. John had no problem at all, he just closed his eyes and went to sleep, an amazing ability he had. The sleeping pill was enough for me so sleep was pretty good.

Q: During your trip, what could you see apart from the Sun, Earth, and Moon?

Duke: That's about it. The Sun was shining so brightly on the spacecraft we couldn't see any stars out the window. We had a telescope we used for navigation and through the 28 power we could identify the major stars.

Q: Tell us about entering lunar orbit.

Duke: The tracking is all done on Earth through the Manned Spacecraft Tracking Network and they give you information that you feed into the computer so that when you run the programs it will orient the spacecraft in the correct position to ignite the engines and decrease your velocity. When you get to the Moon, the lunar gravity is accelerating the spacecraft so you have to slow down to get into orbit. We had to position our spacecraft in such a manner that when we ignited the engine the velocity vector would be decreased in the right direction to put us in an appropriate orbit which was 60 miles by 160 miles. The burn took about six minutes and it took place in darkness. During the burn you take a peek out the window and you don't see anything but you're hoping you are in the right position and are not going to crash into the Moon. Sure enough, right after our burn was complete, the computer said we were in orbit and almost immediately we came into sunlight on the backside of the Moon. It was spectacular—the typical "2001" impression as you entered orbit for the first time. It was very stark and very spectacular.

Q: How did the photographs from other missions compare to seeing the actual thing?

Duke: The photographs pick up the bareness, the craters and the hills, but it just doesn't have the emotion to it that you have as you watch it in real life. You're actually there looking at it with your own eyes and it's a very emotional time, really. As you glide across this lifeless, barren landscape, there is no sound to it. In the spacecraft, you just hear the hum of the inverters and the electrical system, so it was almost an eerie feeling.

Q: When it cam time for you and John Young to separate Orion from Casper and descend to the surface, talk us through the separation and the problem Casper had.

Duke: That problem with Casper was the major problem we had. Prior to that we had a number of smaller problems which impacted our timeline in our lunar module checkout. One was that a communications antenna didn't work correctly. The reaction control system regulators failed on one system. Then we had some other little problems in the lunar module. An hour before we were to land on the Moon, on the backside of the Moon, Ken Mattingly was preparing for a main engine burn to adjust his orbit from 60 miles by 8 miles up to 60 miles circular. This was a major burn. As he was checking his engine out he noticed that the back-up control system, when turned on, caused the engine to oscillate violently. He thought that the whole thing was going to shake off the spacecraft. He notified us and our first impressions were bitter disappointment, because, with the problem he reported, we knew we were in serious trouble and we wouldn't be able to land on the Moon.

Q: You would need to use the lunar module engine to get you home again?

Duke: We would have to use the lunar module to get us out of orbit and get us home. When you decide that, the landing's gone. We got very, very upset. John was concerned, so we started to rendezvous again. We were about a mile apart. We came around the front side of the Moon in contact with the Earth. They told us to stand by and to get ready to come home. We orbited the Moon another time and by then things weren't looking quite so bad. They said, "We're working on it, stand by, there's still a chance." The next time around they gave us permission to land. We were elated and, six hours behind schedule, we made our landing.

Q: Tell us about what you were doing during the landing phase.

Duke: The landing was, to me, the most exciting part of the whole mission. It is very dynamic; you can only plan to a certain point what you are going to see and do. It was a great relief that, after a six-hour delay, we were going to get a chance at it. My job during the landing phase; which took about twelve minutes from the time we started the engines until touchdown, was to monitor the systems, make sure the computer programs were cycling correctly and make adjustments to the trajectory which were called for the by the landing radar. I was mostly looking out the window. Picking a landing site, that was John's job. In the last few thousand feet, he had the landing site in view. I was calling out the altitude, velocity and remaining fuel information so he wouldn't have to look at the gauges, but could continue to concentrate on the landing site. The closer we got, the more he took over manual control of the spacecraft. By the time we were about 500 feet above the Moon, he was in full manual control of the lunar module, which meant throttle and attitude control. At this time we selected a landing site and I began to look more and more out my window, because I wanted to make sure we weren't landing on a large rock or in a large crater that John couldn't see from his side of the spacecraft. I divided my time between the inside and the outside. Right on schedule a little blue light came on inside which meant that we had made lunar contact. We shut down the engine and dropped down the last few feet. We were elated of course.

Q: I remember listening to you as you described the view out the window soon after landing – you sounded like a kid in a candy store.

Duke: You were right! That's the way we felt, like a little five year old on Christmas morning, at the candy store and everything else rolled into one.

Q*: Who named the craters at your landing site and who/what were they named after?

Duke: John Young and I were responsible for naming the craters and the names we selected were those we thought would be easy to remember. For instance, Palmetto is the nickname of my home state. Dot crater is short for my wife's nickname, Dottie. Cinco crater, cinco is five in Spanish and on the photos we saw five craters on the side of Stone Mountain. There was another crater called Cat crater which is an acronym for my two children, Charles and Tom. One crater was called

Lone Star after Texas which is the Lone Star state. These names are certainly not official. They were strictly to facilitate our conversations with mission control during our explorations. There is a group that is officially designated by the scientific community to name lunar craters. To be quite honest, they got a little hot under the collar when we started naming these craters, but now I think it's all smoothed over.

Q: Did you go outside right away?

Duke: That was the plan but, due to the long delay while they were figuring out what the problem was with the Command Module, that extended our day too long. If we had stayed with the original flight plan and taken our first walk right away, it would have meant a 35-hour day before we could have gotten to sleep. We took our first rest period before we went outside. We had to power down the lunar module, take our suits off, change a lot of the flight plan which required a lot of work. We then put up our hammocks and went to sleep, about six hours after we landed. We slept for eight hours, got suited up again and were ready to go outside.

Q*: You didn't have a great deal of space in the lunar module. How did this affect sleeping arrangements?

Duke: On Apollo missions the crews had two small beta cloth hammocks which we attached to the spacecraft and, after taking off our spacesuits, we just lay down in the hammocks and got a good night's sleep. John's hammock was in the upper portion of the LM. The foot of the hammock attached to the sides of the instrument panel just below the telescope or sextant. The head of the hammock went to the rear of the LM about 18 inches above the top of the ascent engine cover and attached to the rear bulkhead of the ascent stage. Below that on top of the ascent engine cover were our two spacesuits. My hammock was perpendicular to John's and was in the foot well of the LM. My feet were basically in the position where John would stand at the commander's station. The foot of the hammock attached to the left bulkhead of the LM about 6 inches off the floor and the head of the hammock attached to the right bulkhead. My head, as I lay in the hammock was basically where I stood at the LM pilot's station. This was quite comfortable, though it was tight quarters in the LM. I found sleeping in one-sixth gravity not much different from sleeping here on Earth except you don't need to move as much to maintain proper circulation.

Q: Tell me about stepping outside for the first time.

Duke: John was the first out. He stepped down and made a little speech. I couldn't wait; I almost kicked him out the hatch, saying "Hurry up John, get goin'" that kind of attitude. Finally, he got out and made sure that everything was right. I shifted sides in the lunar module – you can only get out from one side. I had to cross over, open the hatch and then go outside. I bounded down the ladder and hopped out onto the Moon. I was ready to go from the very beginning. They give you a few minutes to adapt to 1/6 gravity on the Moon, but I was ready to go as I bounded off the ladder. I don't remember what I said, but I was very excited – something like "Ya-hoo!" I guess.

Q:* During the first 'J' mission – Apollo 15 – the commander carried out a stand-up EVA (SEVA) from the LM on the Moon. Was a SEVA from the LM's upper hatch ever considered for Apollo 16 and, if so, why didn't it get the go-ahead?

Duke: Yes, a stand-up EVA was considered for our mission. We sat down as a crew with the geologists and other scientists and tried to allocate the time as appropriates as we could for the mission. It turned out that a SEVA had very low priorities because of the time it took to prepare for that EVA and the dangers inherent any time you depressurize the spacecraft and stand up on the engine cover to look out the hatch. That was a very complicated procedure. It had been done on Apollo 15 but the returns from that were not as meaningful as when you compare it to the time that it took and, more importantly, the time to go ahead and go outside to collect samples and put out experiments. A SEVA was discarded as a procedure for us on Apollo 16 early on in our planning after Apollo 15 results were studied.

Q: Many people know who the first man on the Moon was, but not many know you were the tenth.

Duke: I could have been the 10,000th on the Moon, it wouldn't have mattered to me. I didn't go to get my name remembered, I went for the thrill of exploration and the thrill of adventure.

Q: Early during the first walk when you set up the flag, you took a picture of John Young jumping off the Moon saluting the U.S. flag. Was that planned at all?

Duke: The picture was planned. Each crewman gets his picture taken saluting the flag. It was totaling surprising to me when I saw John jump up off the Moon. I remember I said, "Atta boy, John! Great Navy salute, give me another one just to make sure." You always have to take two pictures so he jumped up twice. I wasn't about to do that because my balance wasn't nearly as good as John's. I had a little problem when I straightened up, I had a tendency to fall over backwards so I didn't try to jump up at that point.

Q: You did seem to have more trouble that John in keeping your balance and fell over more often. Why do you think that was?

Duke: Due to the fact that I was collecting rocks and doing maneuvers in the suit that were strange to me. In light gravity when you grabbed a rock and it didn't move like you expected it to, that would throw you off balance and the torque from that would make you fall over. I did fall over a lot. Fortunately, most of the time it was on my stomach. I did a big push up and got right back up. The bad part is when you fall on your back; then you've got a problem. Not only is your life support system under you, but the suit is so stiff you can't really roll it over, so John and I had to help each other when we fell over on our backs.

Q: During the first walk, as part of an experiment you had to drill into the Moon – was that a tough job?

Duke: On Apollo 15 it was a very, very difficult job. We had made some changes in the lunar drill so it really wasn't that tough. We had practiced and practiced. The only thing I was worried about was that I couldn't get the drill off the top of the drill stem on each segment. It turned out that, with a little tool we invented to help us, it really wasn't a problem. I was proceeding along well with the drilling of the holes when, you may recall, John stumbled over one of the cables that was supplying electrical power to an experiment and, unfortunately, he ripped the plug loose so we lost power to that experiment and had to abandon it. I was very disappointed because that was one of my major goals, to implant that experiment and get it operating.

Q: How did the failure of this, the heat flow experiment affect you and John Young?

Duke: We were disappointed at first, but we were so busy and had so much else to do. We drove it out of our minds and went on to the next thing. By the end of the EVA, basically, I had forgotten about it. When we got back inside, I told the principle investigator for the experiment that we were really sorry it had happened and we wished we could fix it, but there was just no way.

Q: It looked like you had a lot of fun walking on the Moon – a nice way to mix work with pleasure.

Duke: Really! It didn't feel like work at all. I wish every day of my job had been like those three days we were on the Moon.

Q: During your time on the Moon, you commented 'Footprints on the Moon—I can't believe it!' Did you have trouble accepting where you were and what you were doing?

Duke: Surprisingly, it didn't feel like a dream. I didn't have to convince myself that I was there, once we landed. It all came second nature to us. We were there, we were excited about it, it was the culmination of three years of training and it was all a real, livable experience for us.

Q: What was it like executing the physical tasks you had to perform in 1/6 gravity in the pressure suit?

Duke: The problem was the pressure suit. As it inflated to 3.5 pounds per square inch so we could breathe inside, the suit got very rigid and stiff. It was like being inside a big hard balloon. We had to learn to manipulate it so that we could do the manual tasks we had to on the Moon. Fortunately, it was well designed and we were able to do the manual tasks, although we couldn't bend over at the waist and couldn't do a deep knee bend. The suit was just too stiff for that. The manual tasks were difficult because of the "set" of the suit. The suit took a shape and to move it, we had to exert pressure against the suit to move it. The hardest it seemed to me was the gloves. It was like squeezing a hard rubber ball for seven hours. At the end of the EVA we had cramps in our forearms. It was a hard physical effort to move that suit like we wanted it to.

Q: Did the fine lunar surface dust cause you any problems?

Duke: The dust did result in some problems for us by the second EVA. One of the fenders on our lunar rover came off and as we drove across the Moon, the open tire threw up a rooster tail of dust. It was like being in a dust storm as it showered down on top of us covering our suits, lunar rover and equipment. We did a lot of dusting [off] with our dust brush, which took a little extra time than we planned, but it did not result in a major failure of any equipment.

Q: What was it like riding on the lunar rover?

Duke: The ride was very rough and wild. We had to buckle our seat belt because the thing was so springy and the shock absorbers gave us so much rebound that without a seat belt we felt it would throw us out. The only other thing that was impressive was that the back end kept fish tailing; it was like we were on ice all the time.

Q*: During EVA 1, John Young test drove the lunar rover (LRV) in what became known as the lunar 'grand prix'. You filmed this event?

Duke: Yes, I did. I stood to the right of the LRV with a 16mm camera and recorded the entire 'grand prix' on film.

Q*: The LRV appears to get some pretty rough treatment during the 'grand prix'. This treatment could possibly have damaged the vehicle, thus preventing its use in the later EVA's. Wouldn't it have been more prudent to have the 'grand prix" at the end of the final EVA?

Duke: Well, we weren't really concerned about any dangers to the rover, because the profile that was designed for the 'grand prix' had been approved by the engineers and us as a crew, and really consisted of showing the engineers how the vehicle performed on the lunar surface. No one on Earth had ever seen it driven on the surface since the TV camera was always mounted on the rover. The TV could not be turned on when the rover was in motion because the antenna was not gyro-stabilized. Therefore, the performance of the rover was only described by the crew. The 'grand prix' was designed to be a couple of hundred yards in length with one sharp left turn of 180 degree direction so that the engineers who had designed the rover could actually see it in the lunar environment. The film that I took while John was driving the 'grand prix' turned out to be of exceptional quality and gave them some fine indications of how the rover was performing. We didn't do things that would have damaged the rover. We drove it at full speed, made one sharp turn and drove back so it wasn't a real strenuous activity that was not within the envelope of the LRV's design. It was similar, really, to all the driving we did later on during the mission. The reason EVA-1 was chosen was because that was the time it would work without interfering with the geological exploration timeline. The first EVA consisted of getting the rover off the LM, powering it up, getting all the equipment on it and then placing out the experiments and doing a short exploration. During that time it was selected to do the 'grand prix' since I was already off the vehicle and it wouldn't interfere with any of the geological activities we had.

Q: How do the pictures you took on the lunar surface compare to the actual view?

Duke: If you get a good print of the photos we took, it is accurate. It was mostly gray in color, although some were pure white and some were pure black. The sky, of course, is jet black. I have seen some prints which look really brown and some that look really blue –that's not correct. It was gray in color, shades of gray from a light gray as we looked away from the Sun to dark gray as we looked toward the Sun. The Sun angle had the most impact on the color of the surface.

Q: What types of rocks were you looking for at your landing site?

Duke: The photogeology that was done before the mission indicated that we would have primarily two types of volcanic rock in our landing area. That was not true. We had volcanic rock, but we also had many igneous rocks. I believe the majority of the rocks we brought back were breccias – fragments of different types of rock that were welded together. We were surprised and so were the geologists. It was a good geological site because we brought back materials that had not been seen on the Moon before.

Q: At one of your stops during your third EVA you walked over to a big rock. How big was that rock?

Duke: At the beginning, when we looked at the rock, we didn't think it was very large. The problem on the Moon is depth perception. You're looking at an object that you've never seen before and there is no famil- iar scale. By that I mean you don't have telephone poles or trees or cars with which to judge relative size. As you look at a rock, it looks like a rock. It could be a giant rock far away or it could be a smaller rock close in. We thought it was an average size rock. We went off and we jogged and finally got down there and the thing was enormous. I imagine it was 90 feet across and 45 feet high. It was like a small apartment building towering above us as we finally got down there. I believe it was the largest individual rock anybody saw on the Moon.

Q*: Did you venture to the bottom of any craters?

Duke: Neither John nor I ventured to the bottom of a crater such as Plum or North Ray crater. These craters were too deep for us and the walls too steep. The dangers were considerable at a crater such as North Ray which was almost two hundred feet deep. If you accidentally fell into that crater and survived the fall there was no way to extract yourself as we had no lifelines or the ability to pull one another out. Even a forty to fifty foot crater such as Plum provided a major obstacle. If you got to the bottom of it, climbing out would have been an exceedingly difficult, if not impos- sible, task.

Q: What was the highlight for you during your experience on the Moon?

Duke: I think the highlight was when we were near that large rock in an area called North Ray crater. The terrain was very rough; it was scattered with large boulders. There was this deep crater in front of us which was very dangerous to walk around the side of. That, to me, was the highlight of the EVA.

Q: Was there a low point?

Duke: Yes, one. An hour after we were back at the lunar module for the last time. We were getting ready to go back inside for the final time prior to liftoff. I decided that I was going to set the high jump record on the Moon. I jumped up trying to be cute. John and I had planned to do a little play on the Olympics, we were going to have the Moon Olympics. I was going to do the high jump and I started to jump, and as I did, I fell over backwards from about 3 or 4 feet off the ground. As I was going over, I just knew that was it. I would kill myself because the suit was going to split open and I was going to be dead. It turned out that the suit held and I bounced up. I was very embarrassed but very scared also. That was the low point.

Q*: Your were the first crew to have a drink bag installed inside your spacesuit. Could you describe it?

Duke: The beverage assembly was included because the Apollo 15 crew said they got very thirsty on the Moon during a seven hour EVA. We also had an in-suit food bar. The drink assembly was about the size of a hot water bottle and contained nearly one gallon of liquid such as water or orange juice. It was velcroed to the liquid cooled garment we wore and had a tube that protruded from the top of the bag and came up through the suit neck ring and ended about an inch from your left cheek. By turning your head to the left in the suit, one could grasp the tube then bend it to the right—opening the one-way check valve and then drink the liquid. It worked well in lunar gravity.

(Interviewer's note: The beverage assembly did cause a few problems on Apollo 16. Inadvertent activation of the tilt valve by either the in-suit communications cable or the microphone caused the release of a quantity of orange juice into Duke's helmet during the lunar descent phase. Also, Young installed his beverage assembly after donning his pressure suit prior to the first lunar EVA and was unable to position it properly. This resulted in him being unable to have a drink during EVA 1.)

Q*: When were the lunar samples returned to the LM. After each EVA or at the end of EVA 3?

Duke: We would load up our sample bags and boxes and carry them back inside the LM with us after each EVA. This would assure at least some samples were onboard in case we had to abort and liftoff early. Had we left them outside, there would have been the possibility of an early liftoff with no rocks onboard.

Q: How did the liftoff from the Moon compare to the landing?

Duke: I thought the liftoff was very exciting. It was also dynamic and very quick. We bounced off the Moon a lot quicker than I expected as we accelerated upward. It was probably the second most exciting rocket burn of the whole mission, because of the dynamics of it. The lunar module was oscillating back and forth trying to hold its trajectory, so there was a lot going on. We had a great view out the front window. Our liftoff was right on schedule, the burn was accurate, and we were in orbit where we wanted to be.

Q: What kind of impressions did you get from seeing the Earth rise over the Moon from lunar orbit?

Duke: It was the most beautiful sight I had ever seen. As we sailed from behind the Moon, we would see Earth rise. It was a half-Earth as we looked at it and it was a jewel of blue and white. From the Moon, in my mind's eye, I don't remember seeing any continents. I do have vivid recollections of the white of the snow and the clouds and the crystal blue of the water. This little half-of-a-ball hanging in the blackness of space and as we came around, it would slowly rise above the lunar surface. We could see the brown of the lunar surface and the blackness of space and this little jewel of the Earth.

Q*: Following trans-Earth injection, Ken Mattingly undertook a deep-space EVA. What were your duties during this event?

Duke: Basically, I was a safety observer. Mattingly's duties were to go to the rear of the service module where the Apollo SIMBAY was located. There were two film canisters that were to be retrieved since the service module did not reenter the Earth's atmosphere. To get the film back from the mapping cameras that he had used in lunar orbit, we needed to go back and retrieve that film. Those were his tasks and in Apollo, his umbilical line was a combined lifeline, safety line, oxygen line, and a communications line, all in one EVA bundle. He walked back to the rear end hand-over-hand using some handrails and, as he got out of the CM, I followed right after him. My job was to float in the hatch—anchor myself in the hatch with 90% of my body outside the spacecraft and to tend to his lifeline, to be a safety observer for him and to make sure he wasn't going to knock off any of our reaction control jets accidentally or any of his lifelines were going to get entangled in any of the equipment back there. After that process was completed, he retrieved the film which he brought back inside the LM, took the film from him, and then passed them onto John Young who stayed inside during the entire EVA. We repeated the process twice for the two film canisters. After that was over, there was a biological experiment that was mounted on a three-meter pole attached to the hatch and Ken climbed the flagpole out to the end of this biological experiment. While he was doing that, I was back inside the spacecraft facing the open hatch so that I could monitor his activities and be a safety observer for him at that time.

Q: How did that spacewalk experience impact you?

Duke: By now, the Earth was nearly a new Earth with just a little bit visible. I called it a thin sliver of blue and white. As I rolled around to the left and looked over my left shoulder, there was the enormous Moon, a full Moon about 50,000 miles away. At that instant, I felt like I wasn't a player, but an observer, or the audience of this big play that was going on. The stage was the spacecraft, the Earth was in one wing of the stage and the Moon was in the other. It was like I was just observing it all and not really participating. It was a tremendous feeling of freedom to be outside, floating free of the spacecraft.

Q: All good things must come to an end. Was the increase of the 'g' forces during reentry a shock to your system after several days of weightlessness?

Duke: I would say it was. We entered the atmosphere at almost 39,000 feet per second velocity. We were right on trajectory and right on schedule. As the spacecraft plunged into this thin upper atmosphere, it began to light up outside as we ionized the gasses. As we did that, the 'g' levels began to rise and quickly we were above 7 g's. It felt like there was a big elephant sitting on my chest. Going from zero to 7 ½ g's so quickly was a little surprising but, as a fighter pilot, I was used to g's and adapted quickly. During the time we were above this high-g level, we had no problem seeing the instrument panel or performing the functions we had to do to get us down safely. It just felt like you were being pressed down into the couch and you would wish the big elephant would get off your chest.

Q: What was the splashdown like. A big thud?

Duke: It was more than a big thud. To me, we hit like a ton of bricks. I wasn't really prepared for splashdown. I was excited about being so close to target. The aircraft carrier Ticonderoga had us in sight and the helicopters were airborne. I was up on my elbows trying to peek out the right window. I wanted to be in the correct position when we landed so I was listening to Ken Mattingly who was calling out the altitude…500 feet, 400 feet, 300… I said at 200 I'll put my head back and get into position and brace myself. He called 200 and before I could move he called 100 and when he called 100 we hit the water. I had a little whiplash, my head went back and hit the couch strut and all I saw were stars for a little bit; it almost knocked me unconscious. Before I could jettison the parachutes, we flipped over upside down in what was called a stable 2 position. It was going to stay there until we righted it (by inflating some balloons in the nose of the command module). By this time I was looking out the window under water. Once we inflated some balloons the spacecraft flipped back over almost instantly. Forty-five minutes after splashdown we were on the carrier, the end of a great 11-day adventure.

About the Interviewers

Randy Attwood is the co-author of "Moonlander", a book on the history of the Apollo Lunar Module with Paul Fjeld that was published in 1996. Keith T. Wilson is a freelance spaceflight writer based in Scotland specializing in the U.S. space program.

5

"Life Without Dad":
An Interview with
Scott Grissom

Oldest son of Astronaut Virgil "Gus" Grissom
Pilot, Liberty Bell 7
Command Pilot, Gemini 3
Backup Command Pilot, Gemini 6
Command Pilot, A-204 (Apollo 1)

2001 marked the 40th Anniversary of the second U.S. manned space-flight. On July 21, 1961, Virgil "Gus" Grissom climbed into his Mercury capsule "Liberty Bell 7" to ride the second manned Redstone Rocket (MR-4) to a suborbital flight patterned after that of Alan Shepard who flew some two months earlier. The media has often overlooked the many accomplishments of Gus in the wake of his ill-fated splashdown in which his capsule quickly sank after the hatch accidentally blew. To this day no one really knows what happened to that hatch but NASA assured a skeptical public of this extraordinary man's abilities by assigning him to command the first manned Gemini mission with John Young some four years later. Tragically, this most beloved of the original 'star voyagers' was killed in 1967 along with his fellow crew members Roger Chaffee and Edward White, during a routine terminal countdown demonstration test of the new Apollo Command Module in preparation for its maiden flight scheduled for later that same year. The Apollo 1 fire has often been overshadowed by the Challenger Accident. It is our hope that the contributions these three men have made nearly thirty years ago will not be overlooked in our efforts to chronicle the pathway to the stars. In this interview with Gus Grissom's oldest son Scott, he shares with us his memories of growing up as the son of an astronaut and his life during and following the Apollo 1 accident.

***

Q: Your dad was a career military officer prior to being selected as an astronaut. In both cases, the demands on him were great, resulting in his often being away for prolonged periods of time. Having said that when he was home, did he spend 'quality' time with your brother Mark and your mom? If so, what kind of things did you do?

Scott: Well, when my father was [home], which was very seldom, I mean, he was always gone. I think my mom even kept track. There was one period of time when she only saw him four or five days a month. Anytime he was home, he made sure that we did go out and do something. We went water skiing or snow skiing, fishing, hunting, tinkering with the cars or going to the race tracks.

Q: Can you recall any one-on-one type projects that he worked on with you and Mark that were extra special?

Scott: After we moved to Virginia, my dad got some plans out of "Popular Mechanics" for a little sailboat and we built it right in the garage. We painted it bright yellow and we had parachute cloth for the sail. He spent a lot of time doing that. I remember putting it up on the saw-horses and working on it. It took time to learn to sail it. I remember when Hurricane Donna came up the East coast in 1961, the boat hull actually blew out of our backyard and went to another part of the lake.

My mom and dad took off in the family car in the middle of the eye of this hurricane to the other side of the lake to go get it! Through all that wind, my dad picked up the boat and got behind the car while my mom was driving it and he brought that boat back home. We loved that boat and that boat wasn't going to get away.

Q: When you dad was selected as one of the seven original Mercury Astronauts, what was your reaction?

Scott: That just seemed like a normal evolution, it wasn't any big thing. I think probably the biggest shock to me was how the kids at school reacted to it. I didn't know quite what to do with that. As a family we were a very shy bunch and that was really unsettling.

Q: Did you dad come home and share some of his activities and training with you or did he leave that at work?

Scott: Occasionally he would discuss things with us. He was always bringing home things. He would bring home packages of food, a switch, or films. We saw some hilarious films of the test monkeys doing things back in those days. Things that he wanted us to view because he knew we'd get a big kick out of some of the training that he was doing. It was always interesting to us.

Q: You were eleven years of age at the time your dad was selected to fly MR-4. Can you recall when and where he told you of his selection and what was your reaction?

Scott: He took my brother and me to church. This was about a month before the flight. He just said, "I got a secret I got to tell you and you can't tell anybody." That was it. I know my brother had gone up to my mom later and said he had a secret to tell her. She said, "Okay, what is it?" He whispered to her, "Dad's going to fly the next flight!" I think everybody got a big kick out of that.

Q: Did you and your family ever visit the Cape or Pad 5 prior to MR-4?

Scott: We did go out a time or two. The Cape's Chief Security guy was Charlie Buckley. He and my dad were real fast friends. They were always messing around with Jim Rathman and the racecars and what not. I know we got kind of the inside tour on an occasion or two, and I'm sure we did go by Pad 5.

Q: Did your dad take you and your family to the Cape to do any simulator runs with him?

Scott: We got to go to the "white room" up in St. Louis. That was when they were working on the tail end of the Mercury program and just started the first Gemini training. We got to go to Hanger S and a few things at the Cape. He took my brother and me and we got to fly the Mercury simulator, which really wasn't that big of a thing. It was all strictly instrumentation on the inside, there really wasn't any visuals. We did get to fly the Gemini/Agena docking simulator, which was really neat! I learned a lot about air bearings and things like that. That was fantastic how that worked.

Q: Did you get to watch the launch of your father's Gemini flight from the Cape?

Scott: We did get to go. That was a lot of fun. Probably the biggest thing to us, beside the launch itself, was that Mark and I were the first kids to go into the Control Center. We got to listen in on the conversation with my Dad and John Young about the second orbit and that was worth a lot to me. That was really fantastic.

Q: Prior to your dad's MR-4 flight, he flew chase during the MA-2 unmanned flight and had a bird's eye view of that rocket blowing and filling up "his airspace". Did he ever discuss this close call with you at the time or was it one of those things he didn't want to frighten you with?

Scott: No, I've read other accounts about it. None of that seemed to bother my father or spook him. I never knew it happened and I'm sure that there are a lot of things that occurred that we never knew about. That one never came up.

Q: Did your dad ever discuss with you the potential risks involved or associated with spaceflight or was it just something that was understood?

Scott: I think we were encouraged to ask questions if we were concerned. Since the Cold War was right at its peak in those days, I did ask him, "What are you going to do if a Russian sub comes up to you while you're floating in the water?" He said, "I'd probably just yell out "Comrade" and hop aboard.

Q: What was your reaction to his MR-4 flight and did your dad ever discuss the loss of his spacecraft Liberty Bell 7 with you?

Scott: As a child it was kind of tough because you go to school right after what happened and the kids pick on us IMMEDIATELY! You know, "Your dad lost the capsule" type of thing. Of course, I knew it wasn't true. I even asked my father about it and he said, "Absolutely no way." He said they were checking and testing to find out what did happen and they'd do their best to find out what it was. There wasn't any way he had anything to do with the blowing of that hatch. I believed him and I still do.

Q: Can you recall him discussing some of the post-flight investigative tests he participated in regarding the premature hatch detonation?

Scott: I remember him saying they had him in (a mock-up spacecraft). He was in there swinging his elbows around trying to figure out any possible way that something could have come loose, or any way he possibly could have activated the hatch. They just really couldn't come up with anything.

Q: It is a well known fact that every astronaut that detonated the side egress hatch from within sustained a powder burn, a bruise and a cut as a result of the plunger recoiling—even through a gloved hand. The records indicate that you dad never sustained such an injury and that the spacesuit and the glove substantiated this as well. Having said that, didn't you previously tell me that you examined the gloves and the MR-4 suit for yourself?

Scott: Yes, I thoroughly checked the suit and the gloves and there were no marks on them whatsoever.

Q: What are you personal views on Tom Wolfe's book, "The Right Stuff" and the movie by the same title?

Scott: My personal view of Wolfe's book is that it's just a cheap plagiarism of my mom's book, "Starfall". The only things Tom got right at all on the space program were directly out of my mom's book. Everything else, as far as I am concerned, was messed up. If you read one book and then read the other (and my mom's book was out first), you could see where he got his storyline.

Q: Do you share any belief that it was an outright character assassination on your dad?

Scott: No, I'm not sure he intentionally wanted to do a character assassination. I think that he just didn't bother to do any research. There are parts of Tom Wolfe's book that I liked and really appreciated; especially how the pecking order of the pilots go and how the esprit de corps for the whole group works. Wolfe's book is the only book that I've ever read that really captures that, and from that point, I think it's very good. But as far as accuracy goes, it's poor.

Q: Quite frankly, I resented Tom Wolfe's depiction of your mom, as much as I did his portrait of your dad. In particular when he labeled her as "The Honorable Mrs. Squirming Hatch Blower." I am sure your mother is a very strong and courageous individual. I really admire her courage and for the role she has had to play as both the mother and father figure to you and Mark at the most difficult time of her life.

Scott: If you look up "supportive" in the dictionary, you're going to find a picture of my mom. She has always supported my dad and she supported my brother and me with the things that we wanted to do.

Q: What are your views on Curt Newport's efforts to retrieve Liberty Bell 7?

Scott: I'm wishing Curt success for I'd like to see it (Liberty Bell 7) recovered. I think it would especially be good to put it on display so the kids could see it. I think the kids are the future of the whole program. We have to encourage young minds. I think displaying Liberty Bell 7 would be a very interesting story because it involves technology all the way from the start of manned spaceflight to the technology that has evolved to be able to recover things at great depths.

Q: In a previous conversation, you had mentioned to me an absolutely amazing revelation that I never considered possible with regards to the onboard MR-4 cockpit footage that lies some three miles beneath the ocean aboard Liberty Bell 7. Would you comment on that?

Scott: Curt Newport indicated that the FBI could probably do a complete recovery on the footage taken inside the capsule. I'd be real excited to see that.

Q: The camera ran after splashdown, correct?

Scott: I think Curt said it ran for ten minutes after touchdown and that's just about the same time that the hatch blew.

Q: As I understand it, they had plans to dispose of your dad's MR-4 spacesuit. Could you tell us the story behind that?

Scott: I know my dad got real mad about that and he ended up stealing it! He was furious so he brought it home. We put it in a real nice hanging bag and kept it in the closet. In recent years, my mom started to get a little nervous about it. She's very concerned about some of this historical stuff and felt that if it got out, she was afraid somebody from NASA would come out and want to have it back. She felt that they shouldn't have it back because they were going to destroy it anyway, and so she felt they lost their rights to it at that point. I believe that she did the right thing.

Q: Your mom loaned it out for public viewing at the Astronaut Hall of Fame, correct?

Scott: I was going to take the suit to the Astronaut Hall of Fame with the gloves on the wrong arm piece and tell everybody that was how it was worn. I think my dad probably would have gotten a big kick out of something like that, but we put the gloves on the right arm pieces on the suit and took it there where it is on display today.

Q: At the time Project Gemini rolled around, you and your family moved to Texas. As a child of a military officer and an astronaut, was it tough to move all the time?

Scott: We were a military family and we were pretty used to moving around. I think I had four separate first grade classes. It's always tough leaving, but we were military and that was pretty much how we approached things.

Q: Were you close to any of the other astronauts' kids?

Scott: We lived just south of the main gate at Fort Eustis, Virginia. Deke Slayton lived down the street and Wally Schirra lived a couple of blocks over. Kent Slayton was quite a bit younger and we didn't see him a lot. I saw Marty Schirra on a daily basis because we were in the same class.

Q: Having been thrust into the limelight after your dad's missions, did your family lifestyle change and was privacy a problem?

Scott: If he was a prime crew member there were always more problems. It helped the families of the "Original Seven" as they started getting more astronauts and their families in the area and a wider variety of flights. That was actually great for us because it took some of the pressure off.

Q: During my junior high days, your father helped me on a school science fair project through correspondence and it is something I'll never forget. Did he ever help you on any school projects relating to the space program?

Scott: I can remember when I was in seventh or eighth grade in art class and worked on a Rogallo Wing. I took some 3/ 4" balsa wood and built the basic Rogallo framework from an artist concept photo. My dad got some orange parachute material, the real stuff from the spacecraft, and we covered it with that. For the spacecraft, we used a fairly large, probably 1/12 scale, plastic replica coin bank of the Mercury capsule that one of the banks in Florida or Houston had. We took that bank and dad mixed up a little Plaster of Paris and we took a little funnel and snaked it through the coin slot, on top of where the parachute would be, to put a little mass in the bottom of it because it was just a thin little piece of plastic. After we got it weighted right, I attached all the lines to it and we actually flew it like a kite from our Corvette as we drove down the street! When I got it all finished, I showed it to my art teacher and he went berserk over it. We took it out to the top of the bleachers on the football field and threw it off and it flew real good! It was pretty neat and my dad was impressed with me for making this thing.

Q: Your father must have been very pleased to have commanded the first "true spacecraft" that had the capability to change orbital paths and in doing so, dispelled the "Spam in the Can" notion.

Scott: Yes. That's one thing he made clear to us, "Mercury was a capsule and Gemini was a spacecraft."

Q: At the time your father was selected to command Apollo you were 16 years of age. As training proceeded and the launch data approached, could you personally detect a change in his attitude, say, in comparison to what he was like prior to Liberty Bell 7 or his Gemini 3 flight?

Scott: Since I really didn't get to see dad very much, I can't say that I saw a change in his attitude. I had the distinct premonition, say four to six weeks before the first actually happened, that things weren't right. I can't really put a finger on why, but I had a VERY strong premonition that things were going badly. I am sure I was getting that from conversations with my mom. She knew what was happening and it probably wasn't very good.

Q: I have read that your dad confided in you and stated he was contemplating returning to active Air Force duty and volunteering for Vietnam fighter pilot status. Could you comment on this.

Scott: The heart of my father was as a fighter pilot. A fighter pilot is a fighter pilot and he didn't like having his friends over there getting shot up. I think he felt like, at least, the reason or the rationale in those days, as to why we were in Vietnam was a good reason to be there. I think first and foremost, my father wanted to do the best thing he could for his country in any capacity that he could. I think that he felt that was more important than the space program and at that time, he could be used better as a fighter pilot than in working for NASA.

Q: Has it ever been established as to whether your dad hung the huge Texas lemon on the #204 spacecraft or the spacecraft simulator?

Scott: Well, I know that one of those great big Texas lemons disappeared from the house and was going to hang someplace. As to which one it went to, I can't tell you; but I always thought it was #204. I didn't know about the simulator.

Q: How did you learn of your dad's accident and how did you handle it?

Scott: I hadn't been home from school too long and Jo Schirra came over from next door. She had a ghastly look on her face and I pretty much knew, right there, when she walked in the door. As far as handling it, probably the only thing that I could do is try and not be a pain to my mom, because she had arrangements to make and things to do, and try to be there for my brother. I know how much my mom missed my father all the way through his career. My father worked his butt off all the way from day one, and I know how much she missed him. I really felt terrible for my brother, because he was several years younger than I was and he probably had a whole lot less contact through the years with my dad than I had. I felt like I had gotten the chance to know him better and I really felt terrible for my brother. He (Mark) would have essentially been 12 or 13 at the time of the accident. That's always tough. It's always tough to lose a parent, I don't care what age you are. A couple of teenagers without a father to look up to is pretty tough.

Q: What were your feelings in regards to the initial cover up attempt by NASA, when they stated that there was "no word" from the crew at the time of the Apollo fire?

Scott: Most of that stuff I didn't read. I was really too upset to pay much attention to what was going on. Later, when somebody actually tried to pin the fire on my father —that infuriated me. To think that my father would intentionally set the fire. I don't know how to explain it, it goes beyond belief. That's insane! My father always worked more than 100% to get them into orbit.

Q: Very few people today realize that, had it not been for your father's untimely death, he most likely would have been selected to command the first lunar landing mission. How does that make you feel today?

Scott: I'm extremely proud that he had already been told that the first lunar landing mission would most likely be his flight. It wasn't Apollo 11 in those days; I guess it probably would have been Apollo 5 or so. He was to be the commander on it, and that's where he should have been. He had been doing a great deal of the work and putting in the effort to get there. He should have had the first ride and been the first person to step on the moon.

Q: How did you feel as you watched your dad's friends fly to the moon in his place?

Scott: I was proud for all of them. I was proud when John Young got to go. I was especially proud when Al Shepard got to fly. Pete Conrad and I were really tickled when Deke got to finally fly on the Apollo-Soyuz Test Project. My father said all along that there was no reason for Deke not to fly. That whole heart murmur thing had always infuriated him.

Q: Did you father ever discuss how life would be different after his lunar landing mission?

Scott: I think he realized that after the first moon landing things would probably change and I don't think he really liked that. I think he knew that the first person to step on the moon was going to have a degree of responsibility and things would never be the same again after that. That's kind of one of the disappointing things as far as Neil Armstrong goes. Neil did step on the moon first and I'm not sure that he fulfilled the responsibility that was delegated to him by stepping on the moon first. He's been very shy and very, very reclusive. I think he should have done more to get the space program further along by being an ambassador.

Q: Your dad got the bum rap for the corned beef sandwich that Wally Schirra had taken to John Young who, in turn, smuggled it in his suit pocket aboard the Gemini 3 "Molly Brown." As I understand it, Gus took only one bit of it when John surprisingly offered it to him. Could you comment on this?

Scott: That was an extension of their training ritual. They didn't like taking breaks when they were doing their training in the simulators so they got into the habit of packing their lunch and taking it with them. Instead of getting out and wasting time for a lunch break, they'd just sit there and have their meal and get right back to training. The corned beef thing was just a natural extension.

Q: Did you dad ever talk to you about his vision for the space program and where he would like to see it headed in years to come?

Scott: I'm pretty sure I know where he wanted it to go, because I said something to him that I wanted to be an astronaut too and go to the moon. He told me, "Well, by the time you get to be an astronaut and go to the moon, there'll be Coke machines there!" So that's where he wanted it to go and that's really where it should have gone. It's kind of sad to see how it's evolved. I think we ought to go back to the moon.

Q: What do you think your dad would have thought about the reusable winged shuttle concept employed today?

Scott: I know he was actually involved in the shuttle concept early on because one of the things my dad would do is, we'd sit down and have a discussion of the vehicle dynamics of the spacecraft. I recall very distinctly sitting down with a large sheet of paper and we'd draw out a capsule. We drew out how there was an angle of attack at play at the reentry and how that could influence where the vehicle was going to go on the recovery. He even took that as far along as to what some of the future vehicles were going to be like. He told me how they were kind of arguing how the wings worked, but they were basically useless when you were in space. It was an avenue for recovery and they were kind of doing a benefit analysis of it to see how feasible it would be and where they were going with it. So, we were actually discussing the space shuttle back in '65 when we were talking about angle of attack, doing roll maneuvers, changing the center of pressure and doing all those things so you could actually control reentry. They were already thinking about that back in the Gemini Program.

Q: Your mom wrote in "Starfall" about the hate mail that she received as a result of her suit against North American for sloppy workmanship. Were you and Mark aware of this or was it something she concealed from you? Were there any repercussions from within the astronaut corps?

Scott: I was aware of it, but I really didn't know the extent. I think the general perception was that we were just supposed to take our lumps and press on. I know that Gene Cernan was a little more outspoken than the others (astronauts), as far as his "whispering" how my mom approached suing North American. That was okay, I've always enjoyed Gene. I've flown a lot with Gene and I've got to respect his opinion. He's been good for me, personally and career wise. I've learned a lot from the guy. I think he was wrong and I don't think he understood the problem from the view of a widow. That perspective is what Gene might have needed.

Q: Over the years, there have been numerous plans and attempts to dispose of the Apollo One spacecraft. What are your feelings on this?

Scott: I think the spacecraft should be displayed in public. It's better to be displayed than the memory forgotten. I'm real fed up with how NASA's approach has been to it. It's almost to the point that it never happened. They want to bury it some way. Either they want to bury it in a silo or they want to bury it in a trench out in the ocean someplace. They want to pretend that it never happened. So many times you'll see things relating to the Apollo program and there's never a mention of Apollo One and that's wrong. That's flat wrong! If it hadn't been for my father, Ed White and Roger Chaffee they would have never gotten to the moon. There wouldn't have been anything else!

Q: What would you hope that history would have to say about your father?

Scott: I think it has been said best already, "They would never have gotten to the moon without Gus Grissom." I can't remember which book I read this in but I got a big kick out of that because they would have never got there without him. I think that's how it ought to be recorded.

Q: What would you want people to remember most about your dad?

Scott: I think the most important thing is that he worked real hard and he played real hard. He had a lot of fun. He was full throttle all the time whether it was working or having fun.

Q: Of all the things that your father has ever said to you over the years, what one thing sticks out the most in your mind?

Scott: "Do good work." That goes right back to an early promotional poster that Convair put out quoting my father at a pep rally. I've got it hung in my living room at home and it says, "All I ask is do good work." I think that's the bottom line.

Q: What professional career have you chosen?

Scott: I'm a pilot for Federal Express. I love my job, it's a great job. The hours are kind of bad, but the rest of it is pretty darn good.

Q: What professional career has your brother Mark chosen?

Scott: He actually started out as a pilot also. It turned out for Mark that when the air traffic controllers were fired by Reagan, he switched from corporate flying to being an air traffic controller. So he's involved in aviation and he's a fully qualified pilot. He got an ATB and rated in Lear jets, so he's got a tremendous background.

Q: Did your dad's love for aerospace influence your decision to pursue a career in aviation?

Scott: Yes. That's my first recollection of my father. I knew what my father did and that's what I wanted to do. I wanted to fly.

Q: Would Scott Grissom have liked to have been a test pilot and/or astronaut? If so, why didn't this come to pass?

Scott: Well, first off, why it didn't come to pass is because I'm no where near as smart or sharp as my dad was. I wish I had one tenth of my dad's capabilities. I've made two efforts to get into the Air Force and both times they offered me the back of the F-4. My reaction was "No, I'm a front end guy. I don't want to sit in the back of an airplane. I want to be up front." That would have been great. I did apply several years ago to be a research pilot for NASA, and I was real pleased with the effort (even though) I didn't get the job. Joe Algranti did give me a pat on the back, basically, telling me that I was the closest to a pure civilian pilot ever coming to work there. I felt real good about that. I think if I would have had a little more engineering and science in my degree, I'm sure I would have gotten the job.

Q: Undoubtably, it has been tough growing up without your dad. How would you characterize your relationship with him?

Scott: I loved everything my dad was doing and was involved in. I really wanted to be like him, and I aspired to be like him. But like I said before, I'm not one-tenth as sharp as my father was, but I thrust along and do the best I can with what I've got. That's really what I learned from him, you got to do the best you can.

Q: Your dad loved Indy and Stock car racing, as did a lot of the other astronauts, to the point that he raced them, was part owner and even served in the pitcrew. Can you tell us a few of his, and your, fondest stories?

Scott: I could write a book on all the racing stuff!

Q: How about the time when your father and Gordo served in the pit-crew, fuel splashed on them, and the car burnt to the ground?

Scott: The one incident that you're talking about took place at Atlanta Motor Speedway, just south of Atlanta. It was an Indy car or what we called, Champ cars, in those days (Championship cars.) The cars all used alcohol for fuel and our car's fuel tank had worked loose or a fitting had worked loose. I believe Art Pollard was driving the car at the time and had brought the car in for a pit stop. Jim Rathman, Gordo, Bill Yeager, and my father were the pit crew. Gordo looks up and suddenly he's got a mouth full of fire extinguisher and he doesn't realize that the car is on fire because of the way alcohol burns. That was pretty enlightening when I saw that. I didn't get to go to the pits because I wasn't old enough; I was sitting across from where the pit area was. Of course, that took them out of the race. What no one else knows is that the car probably caught on fire another four or five times after that. When they pushed it out of the pits, it was still wanting to burn real bad.

Q: Racing was not always confined to the race track, but expanded to the sky, the water, and even the city streets. Could you tell us one of your favorite stories from each of these domains?

Scott: There were several on the street story side. In the earlier days when we moved down to the Clear Lake area, I know my dad and Gordo got into a race back from the Cape. This race involved both the air and the ground. The idea was, "We're going to see who gets home first." They flew into Ellington and dad parked his jet and ran to the Corvette and took off down the old Galveston Highway. He escaped the cops in Webster, which was kind of a notorious speed trap, and, of course, Gordo was on his heels. Gordo wasn't so lucky. He got stopped for doing something like 120 mph in a 35 mph, so Gordo lost. I can recall another day. We had a boat shop down south of Tenaha. We were in the Corvette one day going down #146 just as fast as the car would go. We were well over 150 mph. It was getting time to shut it down and turn off #146 where Performance Unlimited was. Well the throttle had stuck wide open on the car. That didn't bother my dad a bit, he just reached up and turned the key off. If you thought about it, there were a lot of things that he could have done wrong. I mean, the easiest is if we would have just clutched it, the motor would have absolutely come undone or, who knows, he could have lost control of the car. There were a lot of things that could have gone wrong. It didn't bother my dad. He just reached up and turned off the key. I'm not sure a lot of people could have done that rolling down the road at 150 mph.

Q: In closing and returning to the subject of your father, is there anything that I haven't touched on that you would like to convey?

Scott: The only thing I would like to add is that if my father had not been killed in January of '67, I think his career would probably have gone pretty much like John Young's had. I think he would still be there to this day, if he could. He would still be on flight status because that's what he wanted to do with his life.

About the Interviewer
Rick Boos is a freelance aerospace writer, lecturer, and historian. He has devoted much of his career to researching the Apollo 1 fire, MR-4 mission, and the history of the launch complexes at Cape Canaveral

6

An Interview with
Jim Lovell

Backup Pilot, Gemini 4
Pilot, Gemini 7
Commander, Gemini 12
Backup Commander, Gemini 9
Command Module Pilot & Navigator, Apollo 8
Backup Commander, Apollo 11
Spacecraft Commander, Apollo 13

In the spring of 1970, the entire world was gripped by a drama unfolding in space. NASA's routine launch of Apollo 13 was hardly news in the eyes of the press, who relegated their coverage to the back pages of the newspapers. But two days and 200,000 miles from Earth, an explosion rocked the fragile Moonbound spacecraft. Commander Jim Lovell and his two fellow astronauts watched in alarm as the cockpit grew dark, the air grew cold and the instruments winked out one by one. What started out as a routine flight to the moon – the country's fifth lunar mission – seemed destined for a tragic end. The world followed this story as it unveiled in real time. As we look back in retrospect, it was through courage and ingenuity, by the astronauts and the support staff on the ground, that led to the dramatic rescue.

***

Q: With Apollo 13, one of humanity's greatest scientific endeavors went eyeball-to-eyeball with one of its oldest superstitions. What are your feelings regarding the significance of the number 13 on the mission's eventual outcome?

Lovell: Before the flight, it had never entered my head. Incidentally, I was the prime crew back-up to Neil Armstrong on Apollo 11 and just after that flight my wife and I took a trip to Italy where they said 13 was a lucky number. Of course there are all sorts of things you could look at such as the fact that we launched at 13:13 pm. Houston time and aborted shortly after 10:13 p.m. on April 13th and lost approximately $13 million worth of scientific equipment. You see, one can do all kinds of things like that. But you'll notice – and NASA will say they are not superstitious—that there has never been another "13" spacecraft and there's been over 66 launches of shuttles now.

Q: Fourteen could have been your lucky number since you were originally scheduled to command Apollo 14. Tell us about that fateful switch in mission assignments.

Lovell: Originally, Deke Slayton, who was Director of Flight Crew Operations and responsible for making each mission's crew assignments, put Alan Shepard on Apollo 13 and their crew was training for that mission. We were training for Apollo 14, but the NASA hierarchy came to Deke and said "Look, we think Shepard needs a little more training." He hadn't even been in space, I mean zero-gravity, since his flight of nearly ten years ago, and that had been just a suborbital flight. So Deke asked me if I would switch flights with him. Of course I had just come off Apollo 11, so I was training there, plus I was the navigator on Apollo 8. I said, "Sure". It would mean that I would be at the moon six months earlier. So I gave Shepard Apollo 14 and he gave me 13. Just recently I saw Al and I said to him, "Al, would you like 13 back again?"

Q: You have been to the Moon twice (Apollo 8 and Apollo 13) but never got to walk on its surface. How does that make you feel?

Lovell: Now that I am comfortably here on Earth, it makes me feel a little frustrated. However, when I look back on the event, of course, it was a matter of survival when I was there – I didn't think about not landing on the Moon. Reflecting back today, I take a more philosophical view. I think that 13 was a triumph in courage and in operations of the crew working together as a team with the flight directors and the controllers—all helping to make a successful return. You know, almost anyone with about six months training probably could have been an astronaut going to the Moon. That's it, if everything worked perfectly, because you have books to follow and all you have to do is know what to follow and just keep doing it. But when something unexpected happens, when failures occurs that are not even planned for, not even trained for, multiple failures which occurred on our flight, and you can still successfully bring the vehicle home, that's something else. I think in that respect I have a sense of pride in the "13" flight.

Q: Focusing for just a moment on the actual Apollo 13 mission. Was there a single moment during the flight that stood out as the most anxious for you personally?

Lovell: The most anxious moment probably came about an hour and a half after the explosion when we analyzed the fact that we lost, or were losing, all of our oxygen. You see, oxygen was used to produce electricity in the fuel cells and we were losing all of our electrical power which would prevent us from using the propulsion system. At the time we didn't have solutions yet about how to get home. That was perhaps the most anxious part of the flight as we were 200,000 miles from Earth

Q: Was there ever a moment during the flight of Apollo 13 when you thought you weren't going to come home?

Lovell: Not exactly, because at the low point, when we didn't know how to get home, we still had the lunar module. Even though it was not designed to support three guys for 90 to 100 hours, it still would support us a little bit longer that the command module. At that point we went into the lunar module to transfer the guidance system and everything else, so we were pretty busy. We didn't worry about how long the LM could support us; we just wanted to start working on it and keep ourselves alive by activating the systems.

Q: The seriousness of the accident, I believe, could best be summed up in your words spoken during the mission in which you said, "I'm afraid this is going to be the last moon mission for a long time." What were your thoughts at the time you said this? At this point, did you believe that you would not be coming home?

Lovell: That remark was not meant for publication. We were inadvertently on "hot mike" (the microphone was stuck open). That thing wandered down to the Control Center in Houston. This, of course, was the first flight that didn't have any kind of time delay between the Control Center and the news media, so the news media got it at the exact same time the Control Center got it. But I really meant what I said. At the time, there was still doubt [about] whether we could even make it home safely. The Moon was getting bigger, but we were getting farther away from the Earth. We were struggling to get home and I knew, based upon previous accidents, what the crisis would be at NASA if we didn't make it. So I looked at Fred Haise and said that this might be the last Moon mission for a long time.

Q: What would you have done if you could not come home—if it looked like there was no chance at all for a successful return?

Lovell: We had decided that we would stay alive as long as possible to radio back to Earth anything of interest. All of this, of course, depended upon our trajectory back home. There was a possibility that we could have intercepted the Earth because we got back on that free return course again. If that were the case and we didn't have any electrical power, we would be a nice fiery meteor over the sky for a few brief seconds, or we would skip off the atmosphere and fly toward the Sun or someplace else. If we missed the atmosphere entirely, we would have followed in an orbit about the Earth which would take us around and back out to the Moon again. Most people thought that since one of our oxygen tanks blew, that we were critically short on air to breathe. We had plenty of oxygen to breathe, more than enough to go back out to the Moon again if we had to. It was oxygen for the fuel cells and the water they generated that we were short on and that made all the difference because water was used to circulate through cooling plates to remove heat from certain critical electronic components.

Q: In retrospect, how critical was the timing of the accident? In other words, if it had occurred earlier, later, or at a different point in the mission timeline, what would have been the outcome?

Lovell: At the time the explosion occurred I looked at my companions and said "This couldn't have happened at a worse time." We were some 200,000 miles from Earth and going in the wrong direction. But I was wrong; the accident couldn't have happened at a better time. If the explosion occurred just after we committed ourselves to go to the Moon, still fairly close to the Earth, but heading away from it at a very high rate of speed at that time, we would not have had enough electrical power in the Lunar Module to go around the Moon and come back home again. If it happened after we had detached ourselves in lunar orbit and taken the LM down to the Moon or while the LM was on the lunar surface or on its way up to dock again with the Command Module, we would not have had enough fuel in the Lunar Module to get back home. So, it happened at a very good time.

Q: If this same type of accident had happened during your Apollo 8 flight, you wouldn't be here talking today, because you didn't have a Lunar Module on that mission.

Lovell: Yes, during Apollo 8, if it had happened at any time throughout the mission, we would have been lost.

Q: During Apollo 8, did you have the same type of oxygen tanks that were onboard Apollo 13?

Lovell: We had the same type of thermostat, which was the initial cause of the accident. They were used in Apollo's 8-12.

Q: So this same accident could have happened during any of those previous flights?

Lovell: Yes, except for the fact that there was a series of events that happened in "13" that set up the accident. This was a classic aircraft type accident where a series of events usually takes place and compound themselves to the point where they finally overcome the pilot of the aircraft. Apollo 13 is the perfect example of this. The 28-volt thermostat, which was the wrong thermostat used in all previous spacecraft, wasn't a factor in the earlier Apollo missions because they had never turned on the heater system long enough to cause it to short out and cause damage. In Apollo 13, our oxygen tank was dropped in the factory prior to our flight. This and the fact that the LOX heater system was used to remove the remaining oxygen during a test which damaged the heater system and no one knew about – all of these incidents collectively came together to cause the accident.

Q: What would have happened to the Apollo program if you guys didn't make it?

Lovell: If we didn't make it, I think that the program would have gone on, but it probably wouldn't have gone on as quickly as before. The fact that we got home again and we had pictures of the damaged service module and could analyze the situation, all of this helped to get us quickly back on track. It helped to get Apollo 14 up a lot quicker than it might have. Even if we didn't get back, I think we would have kept going. But you know there were a lot of people at NASA that didn't want to fly after Apollo 11. They said, "Hey, we did it, we fulfilled Kennedy's promise. It's too risky. Let's quit while we're ahead." To me, talk like that was absolutely ridiculous.

Q: Did the accident of Apollo 13 have any effect on the decision made to terminate the later Apollo flights such as Apollo 18-20?

Lovell: Apollo 20 was canceled before our flight. I believe it was cancelled because they wanted to use the Saturn V for Skylab, which was a good reason for canceling it. Apollo 18 and 19 were cancelled after our mission and I think the reason for that is that they weren't completely funded yet. All the money was already spent up through Apollo 17 and to cancel that mission would have been ridiculous.

Q: Prior to Apollo 11, NASA Administrator Thomas Paine told the crew of the mission that, if they did not make it, he promised they would have the very next flight to try it again. After Apollo 13, what was NASA's position about allowing you to try it again? If given the opportunity to go to the Moon again, would you have taken it?

Lovell: He told that to the crew of Apollo 12 as well, by the way, but he never said anything to the Apollo 13 crew. I guess he figured it was pretty safe by then and, you know, I thought the same thing myself—boy, did I goof! In retrospect, I should have asked Paine if we didn't make this thing, could we have another try. And I thought about saying something after the flight too, but at the time there were 20-some astronauts who had not flown one flight and I had four. My thought was that I ought to quit. I didn't think it was fair for me to fly again so I decided before 13 that this would be the last one, although I told the press beforehand that it would be my last flight, only because I thought it would cap a great career in the space program.

Q: During the flight I understand that, due to the cold, it was difficult to sleep. You mentioned in your book, something about a "thin blanket of body heat" would surround you as you tried to sleep. Could you describe this further?

Lovell: In zero gravity there is no convection–heat does not rise. So if you are very still in zero gravity, the body will heat the air next to you and act as sort of a blanket to keep you a little bit warm. But if you moved at all, you would move the air and you would ruin the blanket.

Q: One of the highlights, if I can call it that, of the flight was the improvised way that the mission controllers came up with to modify the Command Module's square lithium hydroxide canisters to fit the LM's circular opening. That was quite a remarkable modification. After that mission were changes made to standardize such fittings for future flights?

Lovell: No. It would be too much of a job [to redesign it]. They might have put just an extra round canister in the Lunar Module in case they needed it. But you know, the system worked so perfectly, so they could have gone back and used the same thing again and it would be fine. That wasn't, by the way, the flight controllers that came up with that solution; that was the people from crew systems that made that decision. They walked in with their device to mission control and told the flight controllers "Here is the jury-rig and it seems to work in our simulator. See if it works in space."

Q: There were no onboard television transmissions after the accident. Why not?

Lovell: We turned off the television camera because it used the high gain antenna which required additional power. To conserve electricity, we shut it down and used instead the omni antennas. The television [image] would probably never have come on with the omni antennas.

Q: Aquarius was supplied with more oxygen that any previous Lunar Module. In retrospect, if this were not the case, how would it have affected your recovery?

Lovell: It would not have affected our recovery at all. We had more oxygen onboard Aquarius because we were going to stay down on the lunar surface longer. But we had plenty of oxygen with the previous lunar modules to make it home.

Q: What were your feelings after splashdown?

Lovell: What a great relief! We were in a crisis all the way down until the time when the parachutes were open because of the pyrotechnics. We did not know if they were damaged. There was also the concern that the heat shield cracked when the explosion occurred. All these little things were kind of getting to us. When I finally saw the three parachutes blossom out and the spacecraft obviously slow down to a nice parachute descent, then I felt a great relief. When I finally hit the water, and no water leaked into the spacecraft, then I thought that we were really in good shape. And we came in apex up (blunt end landed in the water) the way we were supposed to (on Apollo 8 they landed apex down or upside down in the water).

Q: How serious was Fred Haise's urinary tract infection? How did this affect the recovery efforts? Was he in pretty bad shape when he got back?

Lovell: Yes, he had the chills pretty bad during the last day of the flight but still performed his duties very well. When we got back we stuck him in the dispensary in the sick bay of the ship as soon as we got on deck. Looking back, what he did to cause that problem was kind of foolish. During the mission, he wore the motoman's friend which is what we usually wear when we are out in the lunar suit for several hours on the surface. You've got this thing you could urinate in and it had a bag in which you would collect all the urine. Freddo decided why take time out to go pee, why don't we just wear that thing all the time. So, he wore his for three days straight and, of course, that caused an infection to set in. He was in pretty bad shape for a week and a half to two weeks after we landed.

Q: How did the public receive you during your post recovery tour?

Lovell: First, by the time Apollo 13 flew, most countries were already visited by astronauts. I think the crew of Apollo 11 went to 40 different countries and after they returned, the Apollo 12 guys toured quite a few countries as well. Initially, NASA had no plans for a post-recovery tour for us and then all of a sudden they said, "Oh well, there's five countries we haven't visited yet, let's send the 13 crew there." The five countries that we visited were Iceland, Switzerland, Ireland, Greece, and Malta. They all were really great. Overseas people are really very interested in the space program, very appreciative of what we had done.

Q: Of all the official crew portraits, why was that of Apollo 13 shot in your Sunday best instead of the typical spacesuit attire?

Lovell: That's a good question. You know, that has never been asked of me before. That picture was not taken before the flight but after. If you look closely at Haise, he is very thin and pale as he was still recovering from his illness. The reason it was taken in our dress suits was because we were in the process of getting ready for press conferences and debriefings and at the last minute NASA realized that they needed to get something out because we took off without a mission portrait. As it turned out, we did it with suits. The other "official" crew portrait, of course had Ken Mattingly in it along with Haise and myself. This was before Mattingly had to be replaced because of his exposure to the German measles.

Q: What about the plaque clamped to the LM leg. Was it changed to reflect Mattingly's replacement?

Lovell: Bolted to the LM landing gear was the plaque with Mattingly's name on it. With his last minute replacement by Swigert, they build a whole new plaque and had a device on the back of the plaque so that, when I went down the LM ladder on the Moon, I could snap it on the strut. But as we all know, we never got there. We also never got any moon rocks. We needed weight to go into the command module for the proper guidance of the system prior to reentry. One of the things I took back to Earth with me was the plaque I was supposed to take on the lunar surface.

Q: Other than Apollo 11, your mission patch for Apollo 13 was the only one that did not include the crew names. Why?

Lovell: When we got the patch together, and I'm talking now about Mattingly, Haise, and myself, we were the first to take the science aspect of the flight seriously. The scientists and geologists basically said "Hey, 11 landed and they proved that this was an engineering flight and every-thing worked fine." Apollo 12 was still much of an engineering flight. They wanted to test the guidance system to get near the Surveyor. But the scientists were getting edgy so, we went out and were one of the first ones to really start being a scientific type operation. The idea of that patch was essentially mine. I didn't draw it; I drew the other three patches (Gemini 7, Gemini 12, and Apollo 8). I said we wanted to do something with Apollo. I started out the design of this patch with the idea of the mythi-cal god Apollo, driving his chariot across the sky and dragging the sun with it. We eventually gave this idea to an artist in New York City named Lumen Winter and he eventually came up with the three horse design which symbolized the Apollo but also included the Earth and the Moon. The funny thing is that Winter, prior to making the patch for us, made a large wall mural of horses crossing the sky with the Earth below which is prominently displayed at the St. Regis Hotel in New York City. The hors-es are very similar to the ones on our patch, except that it had a fourth horse falling back and that, ironically, could have been Ken Mattingly who was replaced before our flight. That mural now is in New Jersey someplace. Anyway, we said, "Why put the [mission] name on it?" We decided to eliminate the names and instead put in the Latin "Ex Luna, Scientia" or "From the Moon, Knowledge." I plagiarized this somewhat because it is similar to the Naval Academy's "Ex trident, scientia" which is "From the Sea, Knowledge".

Q: Lastly, in light of the demands that your profession requires and its effect on marriages in the program, I believe that you and your wife, Marilyn, are unusual among the astronaut corps in remaining devoted to one another all these years. To what do you attribute your success as a married couple?

Lovell: I had known Marilyn when she was a freshman in high school and I was a junior. Of course, she is a member of the original wives club which is very small in number at this time. But I think that she has supported me, throughout all these years in what I was trying to do. She never was too demanding and seemed to understand the risks that were involved. Mind you, she spoke her piece when she thought it was necessary, like once she said, "Haven't you had enough flights after four?" And you've got to remember, too, that spaceflight was only part of my test pilot career. I did a lot of test work on the Phantom. Even though we were separated a lot I think we kept the rapport going. It is sort of unfortunate that a lot of the guys, when they left the space program after they had been to the Moon, struggled with deciding what else to do. As a result, many marriages suffered and a lot of them got divorces. But we found that there are more things to do as a couple. We have four grandkids and eight great grandchildren now, so you build yourself a little dynasty like that and why rock the boat.

About the Interviewer
Glen Swanson has held a fascination for space exploration since childhood. His career highlights have included founding *Quest*, the only publication devoted to the history of spaceflight, editing *Countdown*, a monthly magazine covering the space shuttle program, and performing oral histories for the NASA Johnson Space Center. He served as editor of the book, *"Before This Decade is Out: Personal Reflections on the Apollo Program"*, winner of the 1999 Pendleton Prize.

An Interview with Fred Haise

Backup Lunar Module Pilot, Apollo 8
Support Crew, Apollo 9
Backup Lunar Module Pilot, Apollo 11
Lunar Module Pilot, Apollo 13
Pilot, Space Shuttle ALT

Fred Wallace Haise was born in Biloxi, Mississippi on November 14, 1933. He earned an Associate of Arts Degree from Perkinston Junior College in 1952 and went on to the University of Oklahoma with the intent of becoming a newspaper reporter. Dropping out to join the Navy, Haise was an aviation cadet from October 1952 to March 1954, and during the next year and a half, he served as a tactics and all-weather flight instructor in the Advanced Training Command at NAS Kingsville, Texas. He subsequently was assigned as a Marine Corps fighter pilot to VMF-553 and 114 at MCAS Cherry Point, North Carolina, before leaving the Marines in September 1956 to resume his education at the University of

Oklahoma.   While attending school, he flew with the 185th Flight
Interceptor Squadron of the Oklahoma Air National Guard until his grad-
uation in 1959 with a B.S. in Aeronautical Engineering with honors.
Haise then joined the NASA Lewis Research Center during which time
the Air Force recalled him into service with the Ohio Air National
Guard's 164th TAC Fighter Squadron.

In 1966, Haise was selected by NASA as one of nineteen astronauts for
Group 6 while a civilian pilot at NASA Flight Research Center at
Edwards AFB.   As a pilot, he eventually logged 9,100 hours of flying
time in more than 80 types of aircraft.

As an astronaut, he served as support crewman for the first flight of the
Lunar Module on Apollo 9 and back-up Lunar Module Pilot (LMP) for
Apollo 8 and 11. Haise flew as LMP for the Apollo 13 mission intended
to land at the highlands of Fra Mauro until the famous explosion that
aborted that mission. In 1972, he was back-up Mission Commander for
Apollo 16 and, subsequently, Mission Commander of Apollo 19 until its
cancellation.

The following year, Haise was piloting a replica WWII aircraft when he
crashed as was badly burned. After recovering from this accident, Haise
returned to NASA, becoming technical assistant to the Manager of the
Space Shuttle Project from April 1973 to January 1976. Haise was
assigned to command one of two 2-man Space Shuttle Approach and
Landing Tests (ALT) in captive and free flights of the orbiter Enterprise.
Later, he was designated as Mission Commander along with Jack Lousma
as pilot of the third planned Shuttle Orbital Flight Test (OFT-2) that
would rendezvous and rescue the ailing Skylab. Unfortunately, the space
station returned to Earth in 1979, two years before the first Shuttle launch.

Haise resigned from NASA in 1979 and joined Grumman Aerospace
Corporation as a Vice President of Space Programs.

***

Q: The preliminary Lunar EVA plans for Apollo 13 had the mission commander walk half a mile out from the landing site while the LMP photographed him. What was the purpose of this setup?

Haise: That I don't recall. Our mission was to land just to the west of Cone Crater, which Apollo 14 did. Then both of us were going to traverse back up to the ridge line and the edge of Cone Crater to sample without a rover. But, I don't recall the experiment you talked about where the commander was going to walk off to some distance alone.

Q: Were there any plans to traverse into Cone Crater if a shallow section was discovered?

Haise: No. That was a fair size feature and, of course, I doubt with the spacesuit you could ever get down into the crater. We had a strap that had hooks on either end for that eventuality, but it was a limited length to utilize if someone slipped or fell in a crater. You could throw them the lanyard that could tie up to a spacesuit and the other person could help pull them out. Cone Crater was quite a fair size feature.

Q: Each lunar landing crew did something unique on the lunar surface. For example Apollo 12 planned to have a photograph of Conrad and Bean in front of the Surveyor with a self-timing mechanisms on the camera and, of course, Apollo 14 had Shepard play golf on the Moon. What were you and Jim [Lovell] planning to do on the lunar surface?

Haise: We had really nothing special planned on the Moon. We had worked very hard to do a lunar orbit show but it was not a joke thing. We got some of the lunar geologists and others to help us build a script on the history and facts about some of the prominent features. We were going to spend several revolutions to do a talk show about the major features on the Moon. We had prearranged with one of the major networks to handle that in a documentary form following the mission. Of course, we never went into lunar orbit so our script went out the window.

Q: Describe how critical the lighting angle was at Fra Mauro?

Haise: The lighting angle was critical for any landing, not just Fra Mauro. You had to have a fairly shallow angle but not so shallow that the shadows were too long. Yet, at the same time, if you went a little higher with too steep an angle it washed out the terrain. The time of landing based on that sun angle was what backed you up all the way through the mission timeline to fix the launch time.

Q: What was Jack Swigert like?

Haise: Jack was a real easy going guy. He was big and well-built having played guard for the University of Colorado [football team]. He was not as much the care-free bachelor as portrayed in the movie [Apollo 13]. Jack wore whites socks in those days, even with a business suit. But he was a very gregarious fellow. He had never married, so he did have girl-friends but it wasn't quite in the light the movie casts.

Q: How far did you and Jim Lovell go to bat for T.K. Mattingly to retain him on the crew?

Haise: There really wasn't any choice. As a recall, there was a meeting with Dr. Paine, who was head of the agency, and Deke [Slayton]. They just would not let Ken fly based on the facts given to them. We really couldn't cycle another month without doing some extraordinary things with the spacecraft. It had already been filled with propellants and had been wetted, if you will, so the valves couldn't have waited another month because of the corrosive effects. They actually would have had to pull the whole stack back and go in to cut stainless steel lines to drain propellant and reweld them. This would have caused at least a month delay and have been very expensive for the agency. So there was really never a choice.

Q: Did you have total confidence in the LM Aquarius to perform its duties as a lifeboat and the PC+2 burn?

Haise: I knew there was no question of the LM capability when we powered it up since this was done before. Apollo 9 was fully powered up and operational in Earth orbit so we really weren't doing anything that had not been done before. The last two burns that we did manually were different. They were done without a computer or automated help. But they were short burns. One was 21 seconds using the descent engine. The other was about thirty seconds using the little RCS jets. So, you really couldn't get too far off target in that short of a burn time. This was another exaggeration in the movie. They made it look like we gyrated all over the place, when in reality, the maximum deviation in any axis was less than one degree.

Q: There were three options for the PC+2 burn: superfast, moderate and slow. Mission Control chose the slow PC+2 burn. If you, as the crew, could have decided, which burn option would you have chosen?

Haise: We were not part of that discussion. We did talk earlier [with Mission Control] about the thought of a direct return. Again, it never came up to us as a question, but we have challenged that. Just as in the movie, they had concerns about what shape the "mother ship" was in. It turned out to be a good choice because, when we later got a look at the vehicle after separating from the Service Module, we were in awe at the extent of damage shown. It looked like the panel, when it blew off, had flipped or rotated, struck the high gain antenna and hit one side of the engine bell. You see, the rocket engine bell has to be of a certain geometry to be stable. If it gets unstable, it might blow up.

Q: Did you consider donning the entire pressure suit for warmth and insulation from the cold?

Haise: Yes, we did. The problem was that the LM had only two sets of hoses, so only two people could get hooked up. The spacesuit interior was a rubber suit and you have to have airflow. Even in that temperature, you would perspire in the extremities, in your crotch and your feet. We knew you had to come out to go potty, and when you did this you would really freeze to death. With the lithium hydroxide problem, we had to use one set of hoses for the makeshift filter modification so we really had only one set of hoses left to hook up to a suit.

Q: Describe your feelings as you went over to the far side of the Moon for the first and only time.

Haise: Jack and I had cameras out. We knew that this was the only time we could see it so we were glued to the windows and shot a lot of pictures. In fact, we shot some of the best pictures of the program, since our flight path allowed us to cover two very prominent seas on the backside – the Seas of Moscow and Tsiolkovsky.

Q: What is your opinion about adapting to weightlessness?

Haise: I have to say, as Jim said in his book, that zero-g is a great, comfortable place to be. There are no hard pressure points and it showed you can move around easily.

Q: My understanding of the reason why you contracted the urinary tract infection was because you wore the urine collection bag for three days straight.

Haise: We had a Urinary Collection Device (UCD), which is the rubber bag you wear in the suit. We also had the back-up Gemini bags, which were smaller bags. It had a drain so you could drain out of it with a hose. I wore one of these and just left it on during the flight. The problem is that when draining, it doesn't always clear all the urine out of the check valve. I suppose that the trapped urine is a possible place for germs to fester, but we were all cold and tired as well as dehydrated. So all of those things, at least for my body's effects, could have been the cause. I'm sure the germs were there that could have gotten anybody sick, not just me; however, I was the one who came down with the urinary tract infection.

Q: Were there any exchanges between the crew or Mission Control challenging or questioning certain items or procedures? Especially during Glynn Lunney's shift, were there any discussions on the intent of getting the crew back alive or just getting them back.

Haise: No.

Q: How serious was the idea to use the crew's urine to replace coolant water in the LM? Could it have been serviced inside the LM?

Haise: Yes, that's true. Again, it was nothing we got to as a procedure to deal with. It was another "what if" like lots of the "what ifs" the ground was playing with.

Q: For separation from the Lunar Module, the docking probe was stored in the Command Module. Wasn't the means of jettisoning the Aquarius done by pressurizing the tunnel?

Haise: Yes. Again we didn't have the propulsion available in the service module, which normally would have been used to affect the separation. Pressure in the tunnel was used to kick it away. During separation, there was a pretty sharp and pronounced jolt, almost as pronounced as the actual exploration.

Q: After the crew's safe return, did NASA management offer the crew another chance at a lunar landing mission like they did with the crews of Apollo 11 and 12?

Haise: No, we were not offered another chance. I'm not sure the story on 11 and 12 is quite a straight story either.

Q: What are some of your favorite mementos from the Apollo 13 mission?

Haise: At home, I have the filter that I used on the AOT where we lined up the Sun to do the manual burns. Jim aligned on the Earth and pitched till I got the Sun. I have the flashlight I used since we never had the lights on after the power down.

Q: What did you like the most and least about the Movie Apollo 13?

Haise: I guess what I liked most about the film was their attempt to make the technical things very accurate and express them in a way that would be understood and impress the audience. I thought the special effects were very well done. Their construction of the spacecraft was very accurate as well as that of Mission Control. Where the actors had to recite technical jargon, they virtually used the air to ground transcripts. They used the exact words, even though I'm sure they didn't know what they were saying, and they threw the switches roughly in the right place for what they were doing. Those kind of things were very well done. I guess several things bothered me for being wrong. Probably the biggest failing, which they couldn't do anything about was there was a larger set of characters that participated in the actual event. I talked to Bill Paxton after the movie about this and he said there was just a limited number of characters you can develop in a two-hour movie. Behind Mission Control there were contractors and subcontractors, and other parts of NASA that participated. Because of the limitations set in film, they couldn't develop the cast of thousands that participated, the big team that really worked out things for us. That was probably the biggest failing, but it's just the virtue of the limitation of movies.

Q: As CapCom on Apollo 14, you saw Shepard and Mitchell walk on the highlands of Fra Mauro. What were your feelings during that time knowing that you almost made it?

Haise: I volunteered for that role. Normally, I wouldn't have been CapCom but it was because I knew the terrain picture and the traverses. In doing this, I felt that at least I could get a little bit accomplished for all the training I've done; I was just using it in a little different way.

Q: What was ARPS like?

Haise: The Aerospace Research Pilot School was for me, a very enjoyable time. During that time, I was able to fly quite a few different aircraft and learn rudimentary aspects of performance testing and handing qualities of aircraft. The only thing that was hard about it was the photo-panel or oscillograph type data. You ended up in those days spending a lot of hours late at night and weekends reducing data manually to write your reports. I guess the advantage of that is you knew what the test engineers were going to have to do, so it made you more sensitive to what it took to collect good data. I don't know what they're using now, but there was no real-time telemetry in those days.

Q: How did you and the rest of the Astronaut Corps feel about Apollo 9 Mission Commander Jim McDivitt's transition from astronaut to NASA management?

Haise: I don't think there was any feeling at all except probably 'goodness'. At that particular time there really were not very many routes within the agency for astronauts to go that's why many of them left. Of course, that changed after Challenger [the Space Shuttle Challenger Accident]. Even today, there are a fair number of astronauts in management positions. Generally, if you left the Astronaut Office and went into a collateral duty or another role, even on a short term basis, you had lost your place in the lineup. It hurt you in terms of getting back in the flying cycle. Of course, with Jim, he had decided he didn't want to fly again and he was a good technical leader. So, it made good sense for both him and the agency.

Q: Gordon Cooper was the back-up Mission Commander for Apollo 10. If he would have been cycled, he would have had command of the Apollo 13 mission. Do you know why he wasn't cycled?

Haise: No, I don't. If I understand correctly, I think Gordo raced cars. There was an unwritten rule before flight that, if you were in the crew cycle, you weren't supposed to do things in which you might be injured because you might affect the mission schedule. One of the no-nos was skiing and the other one was racing cars or motorcycles.

Q: You were the Apollo 11 back-up LMP. Did you get to fly the Lunar Landing Training Vehicle (LLTV)?

Haise: No, only commanders flew and the only back-up assignment I had as commander was Apollo 16. At that time, I did fly the LLTVs. Actually at that time, we were down to one left. We had two that crashed so we lost two out of the three we had for training vehicles.

Q: It has been noted that the crew for Apollo 11 wasn't close like other crews. Did you notice this when you were back-up to the crew?

Haise: No. The crew was different in a way in their assignments, because the back-up crew that was on Apollo 8 was Neil, Buzz, and I. Mike Collins had been injured and Lovell moved up to replace him on the Apollo 8 prime crew. So, Neil, Buzz, and I were the back-up crew. Buzz trained for the center seat on 8. When they went into Apollo 11, we were not consistent between prime and back-up crew. Buzz stayed in the center seat, Mike Collins took the right seat for launch, even though he was the Command Module Pilot.

Q: For the back-up crew for Apollo 11, it was Jim Lovell as Mission Commander, Bill Anders as CMP, and yourself as LMP.

Haise: Initially, it was Jim Lovell, Bill Anders and myself. But Bill left the program early in that cycle, after a month or two, and Ken Mattingly replaced him.

Q: You were named Mission Commander for Apollo 19. Your crew-mates were Bill Pogue and Jerry Carr before they were transferred to the Skylab Program. What was the proposed mission scenario and the proposed landing site?

Haise: They really did not firm up landing sites for Apollo 18 and 19. They were talking about having one of the landing sites on the lunar far-side although I doubt that would have happened since they would have had to build a satellite to enable communications coverage.

Q: Could they have used a modified subsatellite from the SIMBAY as a relay?

Haise: No. They really would have needed a truly unique communications satellite to put in lunar orbit to get the coverage. They were look-ing at getting to the crater Alphonsus on the frontside. There had been reports, in historical times, of sightings of flashes in that crater. That was one of the proposed landing sites. They had four or five that were on their next prime list. They really hadn't settled down and chosen sites for Apollo 18 or 19, which were both cancelled at the same time.

Q: How much training did your crew do for Apollo 19?

Haise: We were in the first month or six weeks of training, as I recall, when they were cancelled. It was going to be a long cycle before we flew. In fact, as I recall now, they were going to fly an AAP (Apollo Applications Program) mission in-between. It was going to be over a year and a half wait.

Q: Would you agree with NASA management's argument that it was best to terminate the Apollo Program early after the first two lunar landings before they had a fatal disaster?

Haise: It's hard to say now. Obviously, they were short of money so it was a problem. I think the later talk was that they were concerned about a failed mission. But if you're worried about a failed mission, you would never have flown the first one. The first one, in fact, normally has the higher probability of failure. I think it was primarily a budget squeeze and the need to move onto the next thing that caused the termination of the later flights.

Q: You conducted the Approach and Landing Tests of the Enterprise. You were scheduled to command STS-3 for the Skylab rescue mission. Why didn't you stay on the shuttle assignment despite the Skylab rescue mission being scrubbed?

Haise: It was just too long a wait. They got into the tile problems and I figured I would have had to wait two to two and a half years before I would fly. It turned out it would have been three [years] for that flight. I was getting up in age and my eyes were getting weak. I planned a second career in management. I had gone to Harvard Business School along the way and had worked in the Orbiter Program Office. I figured it was time to move on.

About the Interviewer
Don Pealer earned a B.S. in Aerospace Engineering at Boston University in 1988 before becoming a Naval aviator.

An Interview with Jack Lousma

Mission Control CAPCOM, Apollo 13
Pilot, Skylab II
Commander, Shuttle STS-3

It was to have been NASA's third lunar landing. But on April 13, 1970, almost 56 hours and 200,000 miles away from Earth, an onboard explosion crippled the spacecraft and threatened to strand its crew in an "absurd, egg-shaped" orbit...for millennia". The story of Apollo 13 is one told on two fronts: one in heaven and one on earth. Here we examine the drama that unfolded in a windowless room where mission controllers struggled around the clock to rescue a dying spacecraft a quarter of a million miles from home.

Former astronaut Jack Lousma worked at the Mission Control Center at NASA's Manned Spacecraft Center in Houston, Texas as the Capsule Communicator (CapCom) of the White Team during Apollo 13. Lousma also served on the Prime Crew Backup for the Apollo/Soyuz mission. He flew in space twice, as pilot during the second manned Skylab mission, and as Commander of the third Space Shuttle mission.

***

Q: What were the circumstances behind your role as CapCom on the white team during Apollo 13?

Lousma: This was my third support crew mission. I also did Apollo 9 and Apollo 10.

Q: You served as CapCom for these missions as well?

Lousma: No. I had not been a CapCom previously. I had been a lunar module checkout pilot, primarily. The support crews were assigned ahead of time, like the prime crews. Crews were already assigned, both primary and support, to Apollo 11 and Apollo 12. I had decided I had done enough lunar module checkups, and since I was going to be on the support crew for Apollo 13, I asked for something different.

Along with being on the command module support crew, I also took responsibilities to work on checklists, flight plans, and communications. While I wasn't the communicator earlier, I got into communications because the guys on the support crew were going to fill the communications role exclusively this time. I was the communicator for the boost at the Cape. I was basically the launch communicator and did the talking to the crew, the astronauts on board or on station at the Cape, and the launch control center. Of course, when a launch clears the tower, primary control switches over to Houston and the guy's job at the Cape is essentially over. So I did that job down there and then moved up to the CapCom job on Apollo 13.

I worked with the crew a lot in training. Since this was a heavy-duty science mission, we had all these lunar traverses that we were practicing. I spent a lot of time with the crew in training in Hawaii, setting up lunar landscapes and practicing the exact traverses that they were going to make on the surface of the moon. I had practiced my CapCom responsibilities with the crew on several of these training trips. I was assigned to be one of the CapComs during the mission. We did a lot of mission simulations before the flight. We did training in communicating and tried to get all the procedures to work out, as well as help the crew get ready for the flight. With these added duties, the role of CapCom was more than just sitting in the control center every once in a while. There was a lot of preflight planning along with a lot of training sessions beforehand, both at the Johnson Space Center and out in the field

Q: Gene Krantz was Lead Flight Director with your team during Apollo 13. Describe working with Gene. What was he like?

Lousma: I had great respect for Gene. I really enjoyed working with him. He was a very disciplined person, a strong leader, and a very quick thinker who was able to take command.

Q: Do you recall what went through your mind the moment Jim Lovell radioed, "Houston, we have a problem."

Lousma: I remember accepting the situation like many of the other problems that we had simulated during the mission. The only difference, of course, was that this problem was real. I don't remember freezing up, being frightened or in a panic about it al all.

It was a legitimate transmission that said, "We've got a problem." It could have been a big one, it could have been a little one, but it required everyone's attention. I wasn't the only guy that heard it, for that transmission got everybody's attention.

My role as CapCom during that time was to keep the crew informed and to work with the crew to make sure that, in anticipation of their needs, I would get the information or action going that I would want if I were them.

The immediate concern on my part was to try and help solve the problem. Of course, identification of the problem was not immediate in coming because so much of the telemetry had been lost with the explosion. Initially, the source of the problem looked like an instrumentation failure that we had simulated many times before on the ground. However, when the crew started reporting what they observed, it was soon obvious that the problem was more than an instrumentation failure. Something was wrong.

Because of the lack of information coming into the control center, we relied greatly on the crew's observations. I was querying them on what they were observing and taking down all the details on what they were reporting. As I recall, they talked about the fact that the meters showed that they were losing oxygen and they could look out the window and see something was venting. They reported their observations and those observations, when pieced together with what little information the ground had, helped to give all of us a better picture of how grim the situation actually was.

After the explosion, the first initiative by the crew for survival was to get into the lunar module. Fred Haise said that they were preparing the lunar module for habitation and they were going down in the lunar module to activate its systems. There were two main reasons for this, one was so they could survive and two was so they could control the spacecraft. We then had to get control of the two vehicles via the lunar module. In order to do that we were going to have to have the navigation control systems activated and state vector [position, speed, and orientation in three dimensions] transferred so it would know where it was and then activate the thruster system so that it could take control and make sure the autopilot was set up to make that happen.

I should point out that before all this happened, it was evening at the space center, and the crew was giving us all a tour of the lunar module by television. It was a nice tour and very informative. They took their time explaining things and everything seemed to be very routine. Behind us, in the viewing room the families were all assembled. They had taken the time to come in to see this television program that occurred during the translunar coast. There were wives and kids, mothers and fathers, aunts and uncles, brothers and sisters, etc. of the crew who had come.

There was quite a large contingent of family present and they watched this whole television show, enjoying every minute of it. Then it was over and everything seemed real rosy and warm and safe and sound and the families and friends got up and went home. Before they got home, the accident occurred. Hence, by the time they got home, everything was in turmoil. It was like the Mutt and Jeff routine where at one time everything was perfect, the other time everything was bad. The families were apprised of the problem and kept up-to-date on it as much as possible. Over the days that followed we made an extra effort to make sure the families were well informed and had all their questions answered as to what was going on.

Q: Listening to some of the audio tapes of the exchanges between the crew and mission control, I was amazed at the lack of emotion in the voices. In the heat of the moment did you find it difficult to discipline yourself to maintain composure as you spoke with the crew? Did you ever find yourself saying, "Don't sound excited, and don't sound human?"

Lousma: I really didn't get emotional about it. It seemed as though it was a serious problem that had to be solved. It was like other serious problems that we had encountered in airplanes before and it seemed to me that there wasn't a lot of excitement or emotion.

Everybody acted very professionally in approaching this problem without being panic stricken. Of course, this is what all of us were trained to do. Once we had the information as to what was going wrong, we began to tackle the problem based on what we knew. The crew of Apollo 13, as well as the mission control crew, were all highly trained.

I think fear and emotion are a result of not knowing what you are doing. We all had very positive attitudes as if we were going to be successful. I was asked after the mission, "What would you have done if you couldn't have gotten them back alive?" and I couldn't answer that question because it never entered my mind that we wouldn't be successful and that they wouldn't get back alive.

I think it is a positive attitude that keeps you from losing it when something goes wrong because what you need to do is—like anything else in life, if you want to be successful at something—project your mind into that successful frame of reference you're going to be in when you're finished with the job. You just know you're going to do it and it's going to be successful. You are successful because you know you will be. If you hesitate or are tentative about it, that's probably the result you will get, a hesitating, tentative result. I can say that all throughout the NASA program, there was a positive attitude. This was also true when we were training to go to the moon. When we went out with the geologists or had people training us about lunar geology, they would never say, "If you get there, you do this." It was "When you get there, this is what gets done." It was like it was a foregone conclusion. I believe this is the attitude we adopted during this particular emergency, and I think that's the reason that it wasn't emotionally upsetting.

Q: The closest human contact on earth that the crew of Apollo 13 had was through you as CapCom. Did you ever try to relay more to the crew through your voice than what was said and likewise could you discern more from the crew's voice than what was said?

Lousma: I think that from the tone you could tell if there was a little tension in what they said compared to the way they would have said it otherwise. But that was true of others on the radio. It was probably true of myself as well, although I didn't notice it. When you are getting ready for liftoff, the launch director's voice is not as casual as it normally may be during a mission. There is a way people have of saying things and you can usually detect some tension. Periodically, I could detect some concern in the voice traffic coming down during the mission. My job was to do the best I could to keep the crew informed, to make them know what on the ground we felt, that all would be well, and to try and anticipate what they were thinking. I wasn't always good at it and probably occasionally said some things they wish they hadn't heard. We tried to interject some humor during the mission but sometimes our humor was not as humorous to them as it was to us.

Q: Were there any scenarios discussed based on the possibility the crew couldn't make it? What would have been the remaining options for them? Were any of these discussed?

Lousma: Not during my watch. I don't even know about it being discussed behind the scenes on the ground. Like I said before, it never entered my mind that they might not make it, so I probably would have remembered it if there had been that kind of message conveyed.

Q: Was there ever a singular moment during the mission on your watch that stood out as being the most anxious for you personally?

Lousma: I think that we always became more conscious of what was happening when some major event occurred during the mission, particularly midcourse corrections. We wanted to make sure that we could get them in the right attitude to come home, which was a little difficult to do sometimes.

I wasn't there for their reentry, I didn't have that watch, so I couldn't comment on that, but clearly, that transition where you dump the lunar module and you're back in the command module running only on batteries ready to jettison the service module into the atmosphere was a nervous time. Had these events occurred on my watch, I am sure that would have been the major event of their return.

I recall being somewhat anxious during the going out and coming back from the other side of the moon, as well as during the time of making midcourse corrections. I was often trying to figure out a way to lift their spirits a little bit when they were cold and nearly a quarterly of a million miles from home.

The mission was a remarkable experience. We had people come and go to the moon before, but not under those conditions of personal threat. So, it's hard to know sometimes how to say things or what to say, but it's always clear that you have to be sensitive to what they are experiencing when you talk to them. Of course, you do try to detect things in their responses.

Overall, I thought they were a pretty cool group throughout the whole mission. I don't remember ever hearing any kind of panic or any kind of emotional outburst or great emotional concern. They were always very professional and calculating in what they were doing and what they were saying. This, I believe, was as much responsible for their successful return as what we did on the ground.

About the Interviewer
Glen Swanson has held a fascination for space exploration since childhood. His career highlights have included founding *Quest*, the only publication devoted to the history of spaceflight, editing *Countdown*, a monthly magazine covering the space shuttle program, and performing oral histories for the NASA Johnson Space Center. He served as editor of the book, *"Before This Decade is Out: Personal Reflections on the Apollo Program"*, winner of the 1999 Pendleton Prize.

9

An Interview with Guenter Wendt

Pad Leader, Mercury/Gemini/Apollo Program
Nicknamed "Der Pad Führer"
Nicknamed "The Benevolent Dictator"

A former Luftwaffe fighter pilot, he earned these nicknames while serving as Pad Leader during the Mercury, Gemini, and Apollo Programs and still wears them with pride and without apology. As Pad Leader, he held ultimate authority over the prelaunch operations in the White Room atop NASA's launch pads, a place where "if things go sour, they go sour fast." With this authority came responsibility for the lives of the flight crews and pad team members - a duty Wendt shouldered with the utmost dedication. His uncompromising approach to pad operations was equaled only by his infamous sense of humor. The games of Gotcha! between Wendt and the astronauts provide some of the most amusing anecdotes from the space program.

Guenter Wendt was pad leader for the McDonnell Aircraft

Corporation during the Mercury and Gemini Programs, then joined North American Rockwell in the same capacity following the Apollo 1 fire. The following was selected from an interview with Guenter F. Wendt conducted by Donald D. Pealer in Coral Gables, Florida, in June 1995.

***

Q: Tell me about yourself.

Wendt: The first five years I couldn't do anything here because I wasn't a citizen. My parents were divorced. My father was an American. He lived in St. Louis. He talked to Mr. (James S.) McDonnell (Jr., President of McDonnell Aircraft Corporation) at the time about giving me a job. Mr. McDonnell said, "Oh, yeah. Bring him over here. We have a job for him." When I applied, he said, "It's great to have you." But when they started working for the Navy, I couldn't get a clearance because until 1956 the United States didn't have a peace treaty with Germany. I ended up working on all kinds of odd jobs like repairing concrete mixer trucks.

After five years, I became a U.S. citizen and I started working at McDonnell in their design department. In a sense, my job was rather demoralizing because during World War II I flew night fighters for Germany. I had an interesting job but it was kind of a let down. My job required me to do probability analysis. It was kind of demoralizing. It was the same feeling I had during the war. I remember once I had to stack bodies up in Hamburg after a fire raid. It was fifty bodies to a pile. All of us could smell the burnt bodies and ashes. You can never get rid of that feeling from your memory. Anyway, I worked for old man Mr. Mac. He owned the company. We used to call Mr. McDonnell "Mr. Mac." He heard about Project Mercury and he wanted to be in it badly. He boasted, "Man! If I have to bid one dollar, I'm going to get that contract." He got it and that sounded great to me since it was a civilian project.

We had a small contract in 1958 for a boost-glide vehicle called the Alpha Draco. McDonnell spent $5 million on the vehicle. They designed it, built it, flew it, and delivered the data to the Air Force for three vehicles. I was operating the launcher for that program. We had acquired an old Honest John launcher and extended its rail to accommodate the boost-glide vehicle. We test flew three of them in Florida. It was my first exposure to the area. I kind of liked it down here.

Q: How did you get involved in the Mercury Program?

Wendt: When Mercury came along, I talked to the project managers. I said, "Hey, guys. You need me." Ed Peters said, "What do I need you for?" I told him what I could do and of my practical experience. He finally agreed with me and I got started in the project. We split up into two sections. One was the Mercury Atlas and the other was the Mercury Redstone. I took the Redstone section.

Two years before we had the actual hardware, I made a flow plan as to how I thought we were going to process this vehicle and launch it. You have to understand the time was in the late '50s and we launched a lot of missiles here. Our rate of failure was about three out of five launches would blow up. We'd remark, "Hey! Look at that. There goes another nosecone."

I made a flow plan and I got my first rude awakening when working with the systems engineers. There were System Leads in areas like avionics, propulsion, and guidance. Everybody wanted to do everything prior to the actual flight in the last seven minutes.

I said, "Hey, guys. Look at this." I made a cartoon that showed the Redstone standing on the pad with five guys leaning into the little hatch. There was an astronaut standing on the crane hook saying, "Hey, fellows?" I eventually convinced them to go ahead with the original plan.

...I asked, "When do you have to do your component check?"
...They said, "At least one hour before the flight."
...I said, "How about three days before the flight?"
...They responded, "I can't do that!"
...I said, "Okay. Tell me why not?"
...They replied, "I don't know."
...I said, "Okay. Since you don't know, it's now three days."

Finally, I made a flow plan two years before we had the actual hardware and we were ready to launch. I only missed the plan by two days when the actual launch occurred. That wasn't bad for the first time of launching a manned spacecraft.

Q: Tell us some of your experiences in the early Mercury flights that flew the chimpanzees.

Wendt: We started out with an unmanned Mercury before flying it with Ham and Enos. Ham was a friendly little fellow. I used to pick him up and take him to the office. I'd show him to the office head and to the other engineers. I always believed that the more people became involved with the hardware, the better response in their work. I was always stuck with being the escort for the VIP visits. I remember once we had to escort a Congressman before Ham or Enos flew.

...He said, "I want to see the monkeys."
...I said, "You're referring to the chimpanzees."
...He said, "They are monkeys."
...I responded, "No. They really are not." But, I let it be. There's a
   difference between a monkey and a chimpanzee. It was obvious to me
   he didn't know the difference.
...He said, "I want to see them."
...I said, "Okay. Let me find out what their status is." I checked with the
   Air Force guys who were running their training program.
...They said, "Enos just came out of training and Ham is in training. You
   know Enos' disposition."
...I said, "Yeah. I know Enos." Enos was a mean son of a gun. Ham was
   really friendly. I went back to the Congressman to explain the situation.
...I said, "Sir. It would be better if we don't see them because they are
   very irritated when they come out of training."
...He replied, "You just don't want me to see them. Do you?"
...I answered, "Sir. It would not be my place to do that. It's just that you
   never know how they'll react."
...He said, "That's just another excuse from you."
...I said, "Sir. Do you really want to see them?"
...He said, "Yes! I want to see them."
...I said, "Fine!"

I knew what was coming. I let him go ahead of us because I knew Enos. Enos had a nasty habit when he didn't like you. He would hunch down, shit on his hand, and throw it at the unsuspecting visitor. He could hit you from the distance of his cage to the door. That's what happened to the Congressman.

...He said, "I guess I know why you didn't want me to go visit him."
...I said, "Sir. I didn't know that. You never know what will happen when you meet Enos."

Q: What was the general attitude among the engineers and technicians during Project Mercury?

Wendt: In the beginning it was quite a different game. You essentially owned the system if you headed it. Everybody had this "can-do" attitude.

...I can remember saying, "We need the power up tomorrow at 10:00."
...The guy replied, "Well. I don't know. That puts me into a bad spot."
...I said, "Hey, everybody else can. Can you?"
...The guy responded, "You got it!"

He might have to stay all through the night but he had it up by morning. Today, all you get are excuses like "Oh. The paperwork isn't right. I don't have the proper signatures." I would have a hard time today living under that environment. It's not like it used to be. You frequently hear things like "That didn't happen on my shift. Hey! My shift is over. Let the other guy worry about it." That didn't happen back then.

Q: Can you tell me about the extent of a day's work in the Mercury Program?

Wendt: We worked ungodly hours. Twelve hours was an easy workday. I sometimes slept at the Cape (Canaveral, Florida) because it was easier to sleep there than to go home and come back again in seven hours. I tried to get all the people involved because it worked to the advantage of the program. You'd get a lot of mileage out of people that are satisfied with what they're doing and they know what they're doing. We had spent long hours on Mercury.

We were running a spacecraft test and it was already 6 o'clock at night. We still had one hour to go before the test was complete. I called the blockhouse. I said, "Hey, guys. We have two choices. We have another hour to work. I know you're hungry. I know you want something to eat and drink. If we decide to power down, it will take two hours to do so. You could go out and eat. It will take three or four hours for all of us to come back. It will take two hours to power up and complete the test. You're going to have one hour left to sleep. Or I can send out for some ham and cheese sandwiches and some coffee. We can keep on going."

The guys said, "Hey. Let's finish it up." I called Ramone's. I told them to send me one hundred ham and cheese sandwiches and two containers of coffee. I told the technician supervisor to send two technicians down to the South Gate to pick up the order. We had ham and cheese sandwiches. I paid petty cash for it. At that time, I could do things like that. It was as simple as that.

Q: How did you get the reputation of being a dictator on the launchpad?

Wendt: We did Mercury, and I got to prepare the spacecrafts for all of the Mercury flights. After the Ham launch (MR-2), we concentrated on (Alan) Shepard's flight. We ended up doing some long tests after discovering something in the spacecraft was screwed up.

Shepard started complaining. I told him. I said, "Hey. If you don't like it, we'll replace you with somebody else. I got somebody who does it for banana peels." So he got up and threw an ashtray at me. He didn't like my attitude. We had a very close relationship with all of the guys. By the time (John) Glenn's flight came along, I got the name of the Pad Führer. I even have a picture of that.

He said, "Dammit! We know one thing. You're nothing but another Hitler."

I said, "It is actually a very benevolent dictatorship. Disagree and the benevolence changes."

Q: (Recall) some of the experiences in preparing for John Glenn's flight.

Wendt: On John Glenn's flight, I got one dollar bills for each member of the spacecraft team and had them signed by John and (M.) Scott Carpenter, who was his backup. We put them into the spacecraft. After he returned, we retrieved the dollar bills. We mounted them on parchment with numbers. There was one inspector that got ticked off because he wasn't included in on the deal. To make a long story short, I got a note from the Senate that did the investigation on finding out who authorized it. It went from Washington to Kennedy (Space Center) to McDonnell. Finally, it wound up with me. They asked me, "Who authorized it?" I said, "I did."

The next word I hear was from some Senator who said, "Do you know that you jeopardized the whole manned space program? Can you imagine him trying to reenter with dollar bills floating around in the spacecraft?"

...I said, "Senator. Do you know where the dollar bills were stored?"
...He responded, "No. But, it's highly illegal."
...I said, "Let me educate you. To begin with, these dollar bills were tightly rolled up and stuck into a piece of thermal shrink tubing. They went through the vacuum chamber and then were installed with a piece of legal paper with John Glenn's signature on it. They were installed in a wire bundle. The bundle was placed underneath the equipment. Now, will you tell me how the dollar bills could get out of that wire bundle to float around in the spacecraft?"

...He said, "I didn't know that."
...I said, "That's the problem with you guys. You open your mouth when you don't know the facts." I can get very upset with Congressmen.

I usually gave the astronauts' wives a tour of the spacecraft before the crew flew. I remember what Anne Glenn said about her husband's flight. She said, "Guenter. Can you guarantee me a safe return?" I said, "Anne. Anybody who will guarantee a safe return is a liar.

There's no such thing as a guarantee. The only guarantee I can give you is, if I am aware of anything that might prevent him from coming back, I'll stop the flight. I don't have the authority to allow the flight to go but I can stop it."

Q: Tell me about some of the gags pulled by the Mercury astronauts.

Wendt: We use to go to great extents to pull Gotchas. It was started by Wally Schirra . You could play a practical joke but it had to be something that hadn't been done before. It had to be original. You had to make sure it was funny but not insulting. We played a lot of Gotchas. You had individuals in the early days like Gordon Cooper. We pulled a stunt on the pad when the press came out on a tour. We had an easy day and Gordo was the guinea pig for a press tour of about thirty people with cameras. The press included Jack King who was on his first big assignment. He wanted everything to go right. They came out in a big old trailer. Dr. (William K.) Douglas, (NASA suit technician) Joe Schmitt, and Gordo were scheming up a joke before they arrived. I didn't have much to do so I rode down that rickety old elevator on Pad 5 and walked into the trailer. There was that damn Cooper. He said, "Hey! Here's what we're going to do!" He told me what they were planning to do when the press arrived.

I said, "We're going to get fired." He replied, "Are you chicken?"

I said, "I ain't chicken. Fine. We'll do it if you do it." When the press arrived, we opened the trailer doors. Joe and I got out first. We walked over to the elevator, stepped in and turned immediately around. Gordo came out followed by Douglas. Gordo very dramatically walked up to the elevator with the portable air conditioner unit in his hand. He stopped and put the unit down. He looked up and down the booster. Meanwhile, Jack King was grinning from ear to ear with all the cameras rolling. All of a sudden, Gordo grabbed on to the side of the elevator and yelled, "Oh, no! I ain't going!" Joe and I reached out and pulled him into the elevator. We shoved (him) right into the elevator. We closed the door and we went up. We went all the way up to the top level. We could hear Jack yelling, "You can't do that! Can that! I want all the film!" The following week in the editorial of Aviation Week there was a reaction to our prank. It felt that Gordo and Col. Bill Douglas, who were both in the Air Force, should be demoted. It demanded that Joe Schmitt and I should be fired because we didn't take the space program seriously. Ever since that day, we'd see some cartoons where Gordon Cooper was depicted as a private fifth class. These were some of things we used to pull.

Q: Describe John Yardley.

Wendt: One of the greatest thinkers that I worked for at McDonnell was John Yardley. Later, he became the number-three man at NASA. John was an excellent teacher. He said, "All right. The way I operate is if you are my System Lead then you own the system. I will not second-guess you. You have complete control over your people and over what you do. If you don't hack it, I'll fire you. It's as simple as that." I liked it that way. In other words, I cannot have the responsibility without having the authority. That showed up several times when NASA picked on John.

There was one time where Page came over to talk with John one on one. He said, "Hey, guy. Your leads have said that they wouldn't work this weekend."

...John said, "Okay. Why aren't they?"
...Page replied, "Because they worked the last four weekends."
...He said, "All right. What did the System Lead Supervisor tell you?"
...Page said, "He isn't going to work his people, either. Hey! I want these people to work."
...John said, "You heard his answer. That is my answer. Now, if you want to call the Old Man in St. Louis, I'll dial it for you."

This is how John Yardley backed you up. I liked that kind of operation. This was more or less a time where we learned things. There were times when I was tagged as a dictator. But, it had worked for me . . .

Q: Describe NASA's relationship with aerospace contractors.

Wendt: It used to be in the old days if a guy needed a tap and we didn't have one in the stockroom, I could give the guy ten bucks to buy one from the local hardware store. That's how we used to do things. I remember we could go over to Martin Marietta to ask for some things we needed.

...I asked, "Do you guys have some one-half inch stainless steel fittings?"
...The guy said, "Yeah. I got some."
...I responded, "Okay. Can I get six of them from you?"
...He answered, "Sure. Just send a guy over to pick them up."

I sent a guy over and we would have our stainless steel fittings.
...The guy asked, "Do you have some sheet metal? I need some."
...He said, "What kind do you need?"
...The guy said, "About 40 thousandth."

...The Martin guy replied, "Yeah. I got some. How much do you need?"
...The guy said, "Half a sheet."
...He'd said, "Okay. Come and get your half a sheet."

We didn't have all the paperwork. We didn't write [up] some cost comparison. We just got things done.

Q: Did you receive much criticism for not following the rules set by contractors and NASA management?

Wendt: I remember the Vice-President of McDonnell asked my wife about my habit. He said, "Your husband does a good job. Can't you make him follow the rules?" She said, "You're asking me?"

He came over to talk to me. He said, "Why don't you follow the rules?"
...I said, "Walter. It's simple. Let me do my job and I'll follow the rules. If not, I'll do whatever it takes to do the job."
...He asked, "What if everybody else does that?"
...I said, "Hey! It's simple. You just change the rules."

I had somewhat of a reputation of not following the rules. I would bend the rules when I felt it was necessary.

Q: Any difficulties with loading or unloading crews from the Gemini spacecraft?

Wendt: That wasn't that bad. They actually had a pretty big hatch and it was easy to close or open. We had a quick opening hatch on Gemini. It required only one hand to open up the hatch. The only thing we had to make damn sure of was that the crew wouldn't pull the ejection handle because we would go up with the White Room. In the Gemini spacecraft, we had to pull seven safety pins for the pyrotechnics. If one pin was left in, it could result in one astronaut being unable to eject while the other guy left.

Q: Was it difficult to keep up with the pre-launch activities with respect to the timeline?

Wendt: There were many things we had to concentrate on doing. If we weren't behind the timeline, we had exactly two minutes to exchange our little presents. There were moments when we were pressed for time. We were always fighting that damn monster in the corner known as the countdown clock. We lived by and died by the countdown clock. Our activities never caused a delay in the flight.

We thought about a lot of things the day before the launch. For example, our six-man team would go out to the pad in two station wagons. We had two cars in case one car would break down. If the car had a flat tire, you would continue to drive on the flat tire. If the car died at the pad, we would have to drag it off ourselves. Nobody could help us. The nearest help was 3.5 miles away. I had a big old tool kit in the car in case of emergencies. We had to think of so many contingencies. We didn't have much time to react emotionally. Everything had to be done at a certain time. It was pretty hectic.

Q: Tell us about the smuggled corned beef sandwich on Gemini 3.

Wendt: John Young. He pretty much pulled that off all on his own. Let's put it this way. I didn't want to hear about it. I didn't want to know anything about it because certain things could bring about some criticism .

Q: Describe the relationship between you and the Gemini astronauts.

Wendt: (Neil) Armstrong asked Pete Conrad about me.
...He said, "Hey! How do you get along with this guy?"
...Pete said, "Hell, that's simple. Just do whatever he tells you."
...Armstrong retorted, "What the hell can he do if I throw a switch?"
...Pete answered, "Nothing. He just steps on your fingers!"

It wasn't that bad. You didn't do anything without my approval. If there were differences in opinion between the technical supervisors and me about doing something, it was resolved by me saying, "You will not do that."

Q: Name some memorable Gotchas during the Gemini Program.

Wendt: I remember a Gotcha that Pete Conrad pulled on me during the Gemini Program. One morning I show up at my office and found the whole room filled with astronauts just standing around. That was unusual because normally these guys did their job and flew back to Houston or went back to their quarters. I had a funny feeling. I thought, "Why are these guys here looking very interested? I'm sure there was a reason." I couldn't figure out what was wrong. I got a message to answer my phone. I opened the desk drawer and a four foot long black snake popped out at me. I pulled that phone right off the wall. There was that damn Pete Conrad. He said, "Gotcha!" The previous night there was a black snake in the training building. He caught it and broke into my desk drawer to deposit the snake. He pulled off an original joke. I figured out how I could get even with him. Two days before the flight, we had the typical press conference. Pete was late as usual. We had started the press conference without him. He came running up on the stage in full view of the press. He started to put on his coat that I had handed him. That's when he found out that the coat sleeves were sewn shut. It looked good on television.

After the Gemini Program, I was given that big check (displayed around the door). It was supposedly worth one million German marks. It was my unemployment compensation because both Mercury and Gemini were built by McDonnell, but Apollo was built by North American. I was working for McDonnell. At the last flight they gave me that check.

In the Mercury and Gemini Programs, I would arrange a dinner for the launch crew after the flight crew returned from their flight. They would tell us what really happened on the flight. We would charter a place and keep out all of the press. I even hired guards. Only the launch crewmembers and their wives were permitted to attend. No requests for autographs or pictures were allowed at this event. The flight crew would open up and tell us what really happened on the flight. We heard some interesting stories.

We had a dinner at the Starlight, which is now a skating rink on Merritt Island. It was a nice dinner. I can recall we had two deputies walk up to Jim Lovell and Buzz Aldrin after they were done with their talk. They placed them under arrest. There was a dead silence in the room.

...Lovell said, "Officer, what is the charge?"
...One of the deputies said, "Passing a worthless check. There's a

gentleman in the back room who said that he has a check that you signed that's worthless." He was referring to the big check the crew gave me as severance pay.

There I was with that big check. I said, "I'd like to cash it." We used to pull things like that.

Q: Discuss your transition to North American and the early work on the Apollo Program.

Wendt: The astronauts asked me to go over to North American. I talked with the people. They were a very old and competent manufacturing company. They didn't agree with my conditions. I told them that I needed complete hiring and firing authority. They said, "No-no. We have people that have twenty years in the business. No newcomer is going to tell them what to do." I said, "All right. I'm sorry. There's no use for me even to think about it."

Then we had that Apollo fire. I was working for McDonnell during that time. It got to all of us quite a bit. We were very close to Gus Grissom and Ed White. By some fate, I was spared from being in charge of it. One of the most unfair questions I was asked is, "Would it have happened if you had been there?" I never will answer that question. I know in my mind which way it would have gone but I won't answer that question. After the fire, I got a call from Deke Slayton who was running the Astronaut Office.

...He said, "We want you to switch over to North American. We have to pull it together or we'll lose the Apollo Program. They're going to cancel the damn thing if we have another failure."
...I told Deke, "Hey, Deke! It's no use. I can't do that unless I can do it my way."
...He said, "Oh. I got a guy here on the phone that says you can do that."
...I said, "Fine. Put him on."

The guy introduced himself. He said, "I'm Mr. (William B.) Bergen. I think our company is willing to meet all the requirements. Deke will explain. Go ahead and talk with the Director."

I went over to the Cape to talk to the Director. I outlined the way I would organize it. He said, "Okay. Fine." He agreed to everything.

I said, "By the way, who's that guy Bergen that called me in Houston?"

He showed me an organizational chart on the wall with pictures. It was essentially a pyramid as I began to look at the top of the chart. On the very top it read, "William Bergen, President of North American."

I told the company before they hired me about the way I made decisions. I said, "You need to understand that I may make a decision based on my conscience. It may cost you a contract or a lot of money. If you don't want that then don't hire me." I had to be free to live with myself. Most people assumed I worked for NASA when actually I worked for North American. The contractor badge didn't mean a damn thing to me. I never worried about that. I had to do my job. The job might be life threatening but I also had to live with myself. I was responsible for the other people working out there on the pad. You cannot base your decision on what is politically correct but rather what is the right thing to do. It didn't faze me a bit. If I thought it was a no-go, they would get the word of no-go. They could fire me. I knew they would be very apprehensive in even considering to fire me. I had built up a reputation by that time. I was very conscious of my reputation.

I changed over to North American. I made some changes. I knew how to do the job only one way. North American had redesigned the spacecraft completely. The ground operation left quite a bit to be desired. I felt we might as well make a clean sweep. I organized a new group. I had to work with volunteers since I couldn't take and assign people to the closeout crew because it was a damn dangerous job. One mistake and we all blow up. Everybody needed to understand that. The whole stack was fueled when we went out to the pad. It had the same power as eight-tenths of an atomic bomb. That's why I always said, "I'll fly higher than you do."

Q: Did you experience any resistance or friction within NASA on how you conducted operations on the pad?

Wendt: It was fortunate for the program that I was able to run the pad the way I wanted. In that game, you don't have much of a chance to form a committee or get an opinion. If things go sour, they go sour fast. And you have to act fast. Many times at night while I was on the river, I would play the old "what if" games. I thought, "If this happens, what can I do?" I realized I would have no chance of asking opinions when the real thing happened.

I remember one time when we were on the Titan pad with 42 to 45 people. We came back after they pressurized the fuel cells with hydrogen and oxygen. I was in the White Room with the one rickety, lousy elevator. I looked at a gauge indicating the concentration of hydrogen gas in the area. I looked at the damn thing and it was at 92%. Hydrogen gas at those levels could easily ignite from a spark generated by brushing your clothes. Hydrogen between 2-92% is highly explosive. It goes through a process called rapid decomposition. It was the same process you saw on Challenger's last flight. Our first thought was "What do we do?" The first thing I did was yell, "Freeze! Do not touch anything. Just stand!" I thought, "Okay. What the hell can I do now? I got to make sure the people don't move because of the possibility of a spark." I heard on the headset people from the ground saying, "What's the problem?" I didn't say anything to them. I said, "Nobody keys the mike!" I realized I had one guy standing close to the elevator. I said to him, "Very slowly. Get a screwdriver and set the elevator lock to open. Then, open the elevator doors." We managed to get that done and the wind blew in. We slowly got the hydrogen level down.

But, this is what I was driving at: I did not have time to form an opinion. I had to act because we could have blown up in a hurry. It was the same with all of the Apollo launches. The flight crew and my closeout crew were up there. We were in the same proximity as the flight crew. We were monitoring the oxygen and hydrogen being loaded onboard. I told the flight crews, "Hey, guys! I'll bet you if we blow up, I'll fly higher than you because you're inside and I'm outside."

These were the realities. We would get into a little power struggle every once in a while. I once had a struggle with Walter Kapryan (a.k.a. Kappy) who was a NASA launch director. We had three ways to evacuate the pad if things went sour. We had a slidewire car. I initiated that concept on Gemini. We had the elevator that would go down to the blastroom underneath the complex. Finally, we could just go down the stairs. They had put a big sign up there that read, "Go slidewire or go elevator." We were in constant communication with the ground. They said, "Copy. Reroute them because that's what the Director wants." It was written that the Launch Director will direct the crew in which way they'll go.

...I said, "Kappy. That ain't going to happen."
...He said, "That's the way it's going to be."
...I said, "No. That ain't the way it's going to be. You can make
    recommendations but I make the final decision because I'm out here.
    It's my tail end. No way!"

He was the customer and I was the contractor, [but] I had a reputation that helped me great a deal. I had the backing of the astronauts. Their support was considered next to God's. I had no problem of convincing them that we had to change the rulebook.

Rocco Petrone was a NASA head honcho during the Apollo era. We had some rules and regulations which were modified during a big thunderstorm. It was lunchtime in the White Room. Everybody was gone except for one guy in the spacecraft, an inspector and myself. Rocco had showed up. We were getting one hell of a thunderstorm. I noticed there was water leaking through the ceiling. I told the inspector to get a roll of plastic and tape. We cut up the plastic and covered the spacecraft. We did a hell of a fast job. I went over to Rocco to explain. I said, "Rocco. You need to understand one thing. You and I just violated about twenty-five rules. I did not have the okay to put the plastic up. I did not have the piece of paper with the proper signatures to allow me to do that. Here's how it would have gone if we went through the official route. I would have called the NASA Test Director (NTD). The NTD would have called the Bendix (Corp.) support contractor. The support contractor would have called the Pad Supervisor. The Pad Supervisor would have had to come up and determine what it would take to stop the leak. He would go down and tell some guys to get some plastic and some tape. In the meantime, we would have had about six inches of water in the spacecraft."

...He said, "I see your point."
...I said, "You're always preaching but it don't work." I had a
    reputation of not following rules exactly to the letter. I always thought
    rules are there as a guideline. I would do whatever was necessary.

Q: How did you keep up with the new developments and knowledge generated by each program?

Wendt: I owed the program to bring myself up to speed in all areas. I took 42 extension courses because in the beginning we knew very little about such things like hypergolics and cryogenics. We were going to work with cryogenic fluids at -425 degrees Fahrenheit. We never did that before. They were all high-pressure gases at 10,000 psi. It wasn't too long before questions began to surface. How do you handle it? How do you pressurize the tank to 10,000 psi? I owed it to the program to insure I was up to speed. I also had to have the assurances that my decisions were final. It worked for me. I got on Apollo and contributed to many of the Apollo decisions that were being made.

Q: What was the policy of the pad members for mishaps on the pad during the Apollo Program?

Wendt: On Apollo, I would always tell the crewman that there would be two possible outcomes if there was an explosion in the Command Module while the hatch is closed and it was a condition red. A condition red was where the Control Center did not know what happened. If we had a condition red and I had reason to believe the guys inside were mobile, I would spend an additional 18-20 seconds to open the hatch. We would all go out together by the slidewire cart. If I had indications that they were badly injured, I would leave them behind inside the spacecraft. The six closeout team members would evacuate. These were the conditions that everybody knew we would follow.

Conversely, if an accident happened while the hatch was still open and some of us were injured, the fire rescue crew would come up and rescue the crew first. These were the conditions or priorities set up. Everybody understood what was going on.

Q: How tedious of a process was it to load the crews into the spacecraft?

Wendt: It actually wasn't that bad with Apollo. You must realize that the astronauts wore these bulky suits. Let's take a guy about the size of Jim Lovell. You put in one astronaut at a time. I always had one astronaut on my Pad Team. He was usually a backup crewmember. He set the switches before the prime crew showed up for loading. We had about 312 steps to complete before we started the closeout procedures.

We would start loading the crew one at a time while the backup crewmember was underneath the seats and standing at the footside of the seats. We had Joe Schmitt or another suit technician get in and hook them up to the air and communications. He would also strap them in. I always wanted the straps pretty tight. In fact, one of the crewmen thought he would get one over on me by loosening the straps after I closed the hatch. He thought he really pulled one on me until the main engines cut off. He almost slammed his head into the forward instrument panel. He realized why I was insistent of having the straps very tight. Once we had loaded the crew into the spacecraft, we had to make sure the switchguard was removed. We had to make sure everything that wasn't supposed to be on the flight was removed from the spacecraft. If everybody acknowledged that the spacecraft was clear, we had the backup crewman crawl underneath the couches to get out of the spacecraft. It could be tough for him to do when we had other packages underneath the couches. Once the backup crewman was out, I received the okay to close the hatch. I would take one last look to make sure everything was removed. I would tell my technicians to close the hatch. We would run a cabin leak check after the hatch was secured. We would then close the outer Boost Protective Cover (BPC). We would break up the White Room and depart the pad area. We had so many steps to go through. We had to make sure we didn't skip a step.

Q: What qualities did you notice about the Apollo crews?

Wendt: Astronauts are very much individuals. In Apollo, there was always the rotation of the support crew, the backup crew, and the prime crew. You couldn't separate the crews once they were put together. The exception to this was the crew of Apollo 11. They were never a crew. It worked out because Armstrong was very capable.

Q: Can you recall some memorable moments in preparation for the flight of Apollo 11?

Wendt: For Apollo 11 we gave each other little presents in the White Room. This was a meaningful way to break up the monotony. I thought,

"What do we give to the guys of Apollo 11?" My idea was to give something associated with an inside story. For example, when a guy is hon-
ored by a city, he usually gets the key to the city. I made them a key to
the Moon. We always exchanged gifts which were meaningful.
Armstrong asked me what he could take along to the Moon for me. I had
given him a little opal stone to take along for my wife. In order to get it
approved, I had to wrap it up, run it through a vacuum chamber and insure
it was oxygen compatible. It was not as simple as just sticking something
in the spacecraft. Everything had to comply with the restrictions. The
wrapping read, "Vent holes. Inside natural stone." I had given
Armstrong a natural stone to take to the Moon. It couldn't be a man-made
one because of the glue on it. I had to comply with so many restrictions.
The opal stone was made into a ring for my wife. This is how we prepared items to send to the Moon. They had a PPG or bag on their suit leg
for any personal items they wanted to take on the flight. It had to be legal.
They took medallions, rings, chains and other values. We had to make
damn sure we didn't have anything that could outgas because the cabin
still had pure oxygen at 5 psi.

The other big gag was that big fish mounted on the wall. It was for the
most important flight, which was Apollo 11. During the preflight quarantine period for the Apollo Program, we had a very limited number of
people that came in direct contact with the astronauts. The size of this
group was about eighty people. They were known as primary contacts.
They were required to go through a two day physical before being permitted to come into contact with the astronauts. It became ridiculous.
Being in quarantine got to be really monotonous. We could go fishing or
waterskiing. I could go up to the NASA causeway with a boat without
crossing over any public highway. I would drive up to the NASA causeway where the astronauts would be waiting. They would climb down by
ladders onto the boat. We would go skiing or fishing. (Michael) Collins
would always tease me.

...He said, "You're such a great fisherman. You have caught all kinds of
   fish. Why don't you have a big fish on your living room wall?"
...I said, "I don't need that."
...He said, "You don't have a trophy, do you?"
...I answered, "No, I don't."

The day of the launch of Apollo 11 arrived and everything was really tense. We were loading the crew into the spacecraft. I put Armstrong in first. He gave me a little present. I put Aldrin in next. Finally, Mike Collins walked up to the hatch for loading. He came into my nice clean White Room with a long paper bag. I said, "Man. It kind of smells fishy around here." He reached into the bag and out he held that damn thing. It was a trophy trout. The plaque that it was mounted on read, "Trophy Trout for Guenter Wendt." I looked at it and realized it was illegal as hell. It was only eight inches long. One foot was the minimum.

Second, it wasn't cleaned. Finally, it wasn't preserved. He said, "Here's your trophy trout." That made all the national headlines. The thing I remember about it was that it was presented in my clean White Room. What really troubled me was how did they get the illegal fish? I found out that the night before the flight wildlife officers had caught the fish. They put it in a freezer. Joe Schmitt bounded that damn thing on a piece of wood.

Q: Recount some funny moments from the other Apollo flights.

Wendt: We did something for every flight. On the Apollo 13 flight, Jim Lovell always used to complain that he couldn't reach some of the switches. I gave him a stuffed rubber glove on the end of an extender. It was his own personal hand extender. Fred Haise's wife was expecting a baby. I gave him a little rubber doll and a pack of diapers. I said, "Since you don't have nothing to do on the flight to the Moon, you can practice putting diapers on the baby." Jack Swigert was a last minute replacement to the crew. He was trying to find VIP passes for all his friends to attend the launch. I made another stack of VIP passes for him. These were all little inside jokes. People from the outside wouldn't even know what they meant. We always expected something.

We had a very neat little thing on one of the Apollo missions. We had this cartoon that poked fun at Shepard since he was the oldest astronaut on Apollo 14. I'm depicted in the cartoon as saying, "Hey. You're so old that we need to give you a piece of lunar support equipment." I was shown in the cartoon giving him a walking cane.

They were filming that serious Colonel Klink stuff in Miami during this time. Shepard sent somebody down there to pick up that damn helmet for me. He presented me with the helmet. Dr. (Kurt H.) Debus almost got ticked off because it had a swastika on it. We did these jokes to break the tension. We were under so much stress that we needed a way to vent it off. We would go off on the deep end if we didn't.

I remember a good Gotcha during Apollo 17 with Gene Cernan. We were in the White Room. The White Room for Apollo was pretty big. I had my office on one side of the room while the other side was a dressing room for the crew. I had a cold that day so I called the Test Conductor.

I said, "Hey, guys. I'm taking myself off the access list. I will not get close to the astronauts but I still have to run some tests." The crew finally arrived in the White Room. I said, "Hey, guys. Get ready to get in there." My station headset allowed me to listen to two active channels. They were the command channel and sidetone channel. I had two telephones, an alarm system, and a regular phone right next to me. I received a phone call.

...The voice said, "This is Dr. Sam Groom from Houston. I'm a flight surgeon. I understand that you have a cold."
...I said, "Yes. I do."
...He said, "I want you to immediately get off the pad. Leave the White Room."
...I said, "Sir. The only way I take directions is from the Test Conductor. That's my chain of command. That's the only way I follow directions."
...He said, "I'm the flight surgeon. I don't have to take that route."
...I said, "Sir. You have to go through the Test Conductor."
...He replied, "Dammit! You are a contractor. I'm a NASA official."

He really got me pissed off. Finally, I said, "Man, just go to hell!" I hung up. Then I noticed that damn Gene Cernan in the corner with a telephone in his hand. He said, "Gotcha!"

We had a party a couple of days before the flight with the primary contacts at the beach house. When NASA bought the property, they had leveled every house except for a three-bedroom beach house. It was a place where the astronauts and their wives could stay during the quarantine. We could use it to get away from the daily grind at NASA. We had a little going away present for the crew of Apollo 17. The area was typical for the Cape. There were no highways or lighting. It was around 8 o'clock when Gene said, "Hey, guys. I hate to break it up but I have to go to the press conference."

We all walked over to his car with a soda can in hand. He jumped into his Chevy convertible. He started it up and began to accelerate. The wheels just spinned. The Chevy didn't move at all. It appeared that he was stuck in the sand. He put it into reverse and accelerated. The wheels continued to spin without moving anywhere. He was really frustrated. He was supposed to be at the Space Center by 8 o'clock. He tried several times but still couldn't get his car moving.

Somebody said, "Hey, guy! You think you can get to the Moon when you can't even get your car unstuck from the sand."

He got ticked off. He said, "Dammit! You guys can get me to the Moon but you can't get me out of the sand! Can you?"

The exchange went back and forth. He really got ticked off. Finally, one guy said, "Hey, Gene. Why don't you take it off the blocks?" The car was about a few inches off the ground.

He said, "Why don't you ask Guenter?" While they were partying I had jacked his car up with some old 4x4s and an old jack. I said, "Gotcha!"

Q: How satisfied were you with your work at NASA?

Wendt: I actually was fortunate to have a job that I really enjoyed doing while getting paid for it. I used to make quite a few tapes for Radio Europe and the Voice of America during the Cold War. My line was always, "Look. The United States can show you what a democracy can really do. Everybody knows I was fighting on the opposite side of World War II. Now they have the confidence to trust me with the lives of the astronauts. There's nobody standing behind me to second-guess me. There's nobody looking over my shoulder. That's how a democracy is run. The communist nations don't even trust their own people." I eventually received a newspaper clipping from the people in Leipzig, Germany. It read, "I have now prostituted my conscience to the

Warmongers of the West." This gave me an indication that I got under their skin. I was never allowed to get close to the occupied Russian territory. I used to go visit my mother and sister who both lived in Berlin. I had to sign out from the American Consulate when I wanted to travel from West to East. They would give me three hours to call back from Berlin to let them know I made it safely. I asked, "What the hell would happen if I didn't call back and they got me?" They answered, "We can't come and get you. But, at least we would know when you got lost." It was very reassuring.

Q: Discuss your experiences during the Shuttle Approach and Landing Tests (ALT).

Wendt: I did all of the Apollo flights and ASTP (Apollo-Soyuz Test Project). After ASTP, I went to Palmdale (California) to do the Shuttle. I also had the job of clearing the Shuttle flights and getting it back to the processing facility. We wanted to transport the Shuttle through the Air Force site. The Air Force wasn't really too keen in having the Shuttle taxi back past the Air Force office buildings because of the hypergolics. In fact, the picture taken of the team after the first Shuttle Approach and Landing Test showed one guy in a pair of coveralls. That's me. We did the flyby there with six T-38s. That was the only time we did that because somebody from the television networks complained about it. They thought we were being reckless since we could be caught in the jet engine vortices. We were prohibited from doing that anymore.

I was condemned for two years to Palmdale by a general manager. We never saw eye to eye on anything. We had a difference of opinion. One day, he came up to me. He said, "Hey, I noticed something. You always ask the people to do something for you. You're the boss. You don't have to ask them. You can tell them."

I said, "For eighteen years, I always asked people to do things. They work their butt off because they want to do the job, not because they have to do the job. I'll continue to work it my way." We always had differences of opinion in how to handle people. I was also a dictator. I said, "If you're on my team, you'll do what I tell you. If not, get the hell out of here."

Q: What kind of jokes were pulled during the early Shuttle Program?

Wendt: During the Approach and Landing Tests we had ABC and CBS doing a television series out there. I always made Joe Engle and Dick Truly put yellow overboots on before getting into the Shuttle so they wouldn't track in dirt. They said, "Man! They ain't going to see the two of us with these big duck feet on." We made a big cartoon of it showing them going out naked. These guys retaliated the next day with a cartoon depicting us loading ammonia on the Shuttle while they made a flyby fifty feet off the deck in afterburner. These are the things we used to do to break up the monotony. We needed to laugh at ourselves.

Q: Were there major differences in the way things were done in Apollo Program compared to the Shuttle Program?

Wendt: People made decisions more on the technical merits than on the political ones. We had more or less a "Let's get together and talk" session. You had engineers not politicians there. They asked, "What can we do? How can we do that?" It was not a question that we couldn't do it. It was more of an attitude of how we would go about accomplishing the task. I told them, "You got that system. Take a look at it and come back in an hour. Tell me when you can be ready and what do you need to have." They came back and told me what they needed. Decisions were made more on a technical basis rather than a political one. I had fewer people that worked for me back then. There weren't so many people involved in it like there are today. There are now more turf wars. I always used to say that I could best evaluate an engineer not during the normal launch countdown but when something went sour. This forced him to do troubleshooting. That's when I knew if he could hack it or not. He isn't worth a damn if he has to sit there and fumble along. We were technical people. You make decisions on a technical basis rather than an emotional one. That wasn't a big deal because we always played the "what-if" games.

Q: What do you think the problem is today with NASA?

Wendt: I think it is the leadership. The leadership isn't there. The other factor is the Challenger accident. I looked at flight crew safety for Rockwell at the time. We didn't have a problem with the Orbiter. I was spared from being responsible for the Challenger accident. It's funny how fate sometimes plays out life. It would have finished me if I was responsible for that accident. After the Challenger accident, we had all these Congressional investigation committees. It used to be that a piece of legal paper required only a signature from the contractor and the customer. Now we require at least eight signatures on a legal document because everyone assumed someone else looked at the damn thing. In fact, nobody looked at the damn thing. Everybody wanted to avoid the responsibility.

I used to conduct meetings in a certain way to reach a decision. I said, "I need a decision. You're speaking for this company. Can you commit to this decision?"

...The contractor replied, "I'll have to talk to my manager."
...I said, "Sir, I don't want to see you at the next meeting. Send somebody who can make a decision."

Everything today is done by committee. Nobody is responsible for anything. I can't tolerate that.

I flew back to Houston in the same plane with Bob Crippen. We talked. I said, "Hey, Bob. What we need to do is make the people responsible again. Don't make a decision based on six people or a committee. Put the responsibility of a decision on an individual. Fire them if they don't hack it."

Bob said, "You can't do that." People used to admit to making a mistake, but now they are trying to hide it. I'm sure that will lead to another disaster. I already know of instances where things were not in order. I was not actively involved but I know of people who are. I have a pretty good feeling what goes on there. That's what scares me.

NASA tries to play it politically correct. Alarms go off in my head when I hear (NASA Administrator Daniel) Goldin say that NASA will have to cut another $107 million from the Shuttle Program and the Shuttle Managers agree that they can do that without compromising safety. We don't have any leadership.

The Space Station is the next big program for NASA. Who is the head of the Space Station? The Space Station is currently on its eighth manager. This is the problem. You knew who was running Mercury, Gemini, and Apollo. There is a problem when NASA has to assign an eighth manager to the Space Station. They receive an edict to be politically correct. They follow with whatever is being told in Washington.

Before I retired, I was on the accident and incident reporting system. It involved a high level of contracting people. We had the Vice President of Lockheed on it. They looked at various incidents and came up with a solution. For example, we looked at a worker who overpressurized an OMS (orbital maneuvering system) tank. Their corrective action was to give the engineer one week off without pay to penalize him for his mistake. They felt the item was closed.

I said, "This is wrong!" Here was a Rockwell guy telling a Lockheed guy that NASA was wrong. They said, "What do you mean it's wrong?" I replied, "You've taken care of the symptom. How about taking care of the problem?" The head man from Lockheed said, "What do you mean? What is the problem?"

I said, "If your people haven't told you then let me enlighten you. To begin with your engineer was watching two consoles in the Firing Room which is not legal. Second, he was twelve hours on station when he should have been on for only eight hours. Third, he started the pressurization process when he got a call from his manager to explain what he was doing. In the meantime, the pressure ran away from him. If you want to correct the problem, let the people watch only one console, don't let them work for twelve hours, and tell that manager to get out of the office if he wants to talk to them but not during critical tests." After my explanation, I got a look as if I had overstepped my bounds. I used to fight that all the time. The problem is they don't want to owe up to it. That's the reason I acted as the devil's advocate.

Q: Did you find anyone else within NASA management that shared your opinion?

Wendt: Sometimes I couldn't get things out in the open the way I wanted to. I would call Houston and talk with John Young. He would attend the prelaunch review. I said, "Why don't you ask the question on this subject? The answer should be this, this and this." John Young is extremely dry. You always have to wait two minutes after he said something to understand what he really meant. He would come up and ask the questions. The guys would do a toe dance in trying to answer the question. He is technically an extremely capable guy. But he has tough time because he's not politically correct. He will say things that are correct but they are not politically correct for NASA to say.

Three weeks ago we had the Space Congress here on the beach. John was delivering a paper. After his delivery they asked him how he really feels about things. He said, "Let me put it this way. We are going out to colleges and we are telling our young people that they need their education. We are telling them they need a top level education in science and mathematics. We're telling them that by getting an education they are getting a key to the future. Let's say we'll do all this. By the year 2000, we'll have a high level of educated people that can balance their checkbooks. That will be the only job we'll have for them. We cut back research and development, and use their talents to balance their checkbooks." That's John Young for you. It's true but people don't like to hear that type of talk.

Q: What do you think NASA should do to sell the manned space program to the American people?

Wendt: I would propose that we build an orbiting power collecting station of about 100 megawatts. We should get the power transferred down to Earth for distribution and show it can be done. Once we show that we have an unlimited source of power, which is free of pollution and profitable, there will be a lot of big industries that are willing to participate. We can help them along by giving them low cost loans for development of the system. Let them develop it. They will do it if there's money in it. NASA has done a lousy job of publicizing the benefits from space. Everybody wants to know what's in it for us. We need to be able to respond to that.

Q: Do you think there will be light at the end of the tunnel for our space program?

Wendt: I think at the present time the answer is no. The situation could change if we get a different direction or a different commitment. We need to convince people that this is what we want to do and this is why it is good for us. There are a lot of things we could show the people to convince them that this is good for us. That this is good for the children and our future. The only way we can maintain our standard of living is through our science and technology. That's the only thing we can make money on. We need to get our people educated. We need to go in a different direction. A nation that cuts its research and development is going to go down the tubes. We also give ammunition to the opposition if we go ahead in investigating why a billion dollars was spent on the theory of what happened one billion light years away. The average guy asks, "Spending one billion dollars for something that's happening one billion light years away. What does it do for me? I'm not against science but NASA has done things to verify the egocentricity of some scientist that wanted to prove that some particular thing happened one billion years ago. It doesn't really do anything for us.

We need to get the general public involved. We need to do it through an educational process to show what it is about. When I go to lecture to high schools and colleges, I have really nothing for the young people to look forward to in the future. In the beginning, we told the youth that space was the next thing to look forward to in the future. How can we go ahead and encourage our young brilliant minds to go for something that we don't really have. That's our problem because we don't have the correct priorities. I (would) like to have some really dynamic people involved at NASA who could convince people. We need to convince Congress that this is something for everybody to invest in. But, we don't have that.

About the Interviewer
Donald Pealer earned a B.S. in Aerospace Engineering at Boston University in 1988 prior to becoming a Naval aviator.

Gen. Bernard Schriever
Father of the Air Force Space Program

Gen. Bernard Schriever, the recognized father of the U.S. Air Force's space program, is a space pioneer in every sense. Under the tutelage of legendary Air Force General Hap Arnold, Schriever learned the importance of research and development, then applied those concepts to the Cold War. He and a handful of other officers successfully lobbied for the successful integration of R&D into the Air Force's command structure. The results speak for themselves.

As the first commander of the Western Development Division in 1954, he undertook responsibility for Weapon System 117L, the initial Air Force space program. Throughout the 1950s and 1960s, he answered the challenges posed by the Soviet Union and fostered the growth of the nation's new ballistic missile and space programs. He is considered responsible for the development of several space and ballistic missile programs including early satellite reconnaissance efforts and man-in-space research, as well as focussed efforts on the Atlas, Titan, Thor, and

Minuteman missiles and their launch, tracking, and support systems. The manned space program saw Gen. Schriever's guiding hand as the director of the Air Force's Manned Orbiting Laboratory Program and by his lending of staff and expertise to NASA.

***

Q: To begin with, we'll look at the early developments of the American ICBM (Intercontinental Ballistic Missile) Program and the response to the Soviet developments of the ICBM Program and how that affected the American program, if you could tell us about that.

Schriever: Well, I think I need to set the stage a little bit and then I think (things) will go smoother. At the end of World War II, General (Henry H. "Hap") Arnold, who was my mentor and certainly was the most visionary Air Force officer that we had up to then and, as far as I'm concerned, in the history of the Air Force, said that, "The next war will not be like the last one. World War I was won by brawn, in the trenches. World War II was won by logistics," and I can vouch for that, because I spent almost three and a half years in the Pacific (Theater), and logistics was very, very important in winning that war, as well as other aspects. "World War III," he said, "will be won by brains," and he went on further to say that the breakthroughs that really occur, that are most important, were electronics, flow of information, jet propulsion, rocket propulsion.

Let me stop there with rocket propulsion, because the interest in long-range missiles started right at the beginning of the period following World War II, so it has been a long-term interest. We started in the Air Force by building a large rocket facility in California, starting shortly after World War II. We managed to get quite a few of the German scientists who were involved in the V-2 program, and so did the Soviets.

So that's how we really started, and we had a great deal of interest in a long-range missile, but there were other technical problems which didn't really make sense for us to start a full-fledged long-range missile program, but we were certainly well into the program with respect to first the nuclear weapon component of it and other aspects from a technological standpoint. So we started right after World War II. As far as our intelligence indicated, so did the Soviets.

Q: When they did start on the ballistic missile program, was there information that you had about how they (the Soviets) were progressing or what stages they were at, that you could then measure progress against?

Schriever: Well, we didn't start our program until, really in earnest, in 1953, I mean, as far as a weapon system development and acquisition program was concerned. That was after we had actually progressed from the fission nuclear weapon to the thermonuclear, which gave us a much more effective warhead which provided a yield of a megaton with a weight of 1,500 pounds, which is roughly at least an order of magnitude superior, from a weapons standpoint, than was the fission weapon. That really provided the spark to get our weapon system program going in 1953-1954 time period.

You asked about the Soviets. We didn't have really any hard intelligence information about where they were with respect to the ICBM, but our studies indicated, and there was information available, that would lead us to believe that they were very much involved in an ICBM Program. That is about all you can say with respect to the '53-'54 time period. We were concerned that they would beat us to the draw; in other words, of getting a thermonuclear military capability going with long-range rockets. And that's what started us by putting a very high priority on getting on with an ICBM Program of our own.

Q: As the ICBM Program began development, (were) there any discussions or thoughts on not just applying it as a missile, but also applying it to space?

Schriever: Well, here again I hark back to General Arnold. He created the Air Force Scientific Advisory Board, and one of the first things he did after the war was to ask the Scientific Advisory Board, as well as the RAND Corporation, which was set up to support Air Force thinking with respect to the application of technology to the future, what is the feasibility of a reconnaissance satellite. So we were actually working on the idea of reconnaissance satellites starting back in the middle forties, after World War II. Here again, we were involved in technical planning, as well as some testing, but we did not have the capability of putting anything into orbit at that time. But the interest was there and we had always involved in research programs of this type. So our interest was very high, starting at the end of World War II, for a reconnaissance satellite program, and we started on that as well, in the same manner as the missile itself.

Q: When did you learn of Soviet efforts for a satellite program and actually learn of Sputnik as well? And how did that impact your job?

Schriever: Well, that's a long story, but I'll try to make it short. We did approve and provide the highest national priority to the ICBM Program when the Scientific Advisory Board, a special committee of the Advisory Board, recommended that the Air Force do so. That was approved within the Air Force circles and also by the political side, the Secretary of the Air Force, and the Air Force gave it the highest priority as early as 1954. It was approved in the White House, but it was not until 1955 that President (Dwight D.) Eisenhower gave the ICBM Program the highest priority of any weapon program in our inventory, so to speak.

A part of the overall program was to get more hard information with respect to what the Soviets were doing, and one way we achieved that was by establishing radar coverage of where they were doing their testing. That was a capability that gave us a lot of detailed information as to where we stood vis-a-vis the Soviet Union in the development, because we could gauge information from their test flights and so forth. You can't hide an ICBM, you know, when you fire it. So we had information within the next couple of years pretty much on where we stood in connection with the Soviet Union in the ICBM area.

From the standpoint of the satellite for reconnaissance, that was a different matter, and it was highly classified, but we were moving forward on such a program, not quite at the rate that we could have, but we were putting first priority on the ICBM. But the satellite program also was very important from an intelligence standpoint, so that was given quite a bit of emphasis and it was also part of the responsibility of what was called Western Development Division (WDD) on the West Coast, which I commanded for about five years. It was [later called the] Ballistic Missile Division (BMD) and we had the responsibility for the satellite activity of the Air Force.

Q: When Sputnik was launched, what was the general reaction, and when did you actually hear of the launch?

Schriever: Well, let me say it was no surprise. We had the capability of putting something up into space just to prove that we could put a satellite up there, but we were not given the authority to do that. [It was] the IGY (International Geophysical Year) and a scientific endeavor [was proposed but] the [Vanguard] launch was unsuccessful. The Navy and NASA were involved in that first launch.

But the Sputnik did one thing that was very much a plus; it woke us up and it concerned the American people very much that they beat us to the draw in getting the first satellite into orbit. But we at our level, with the information that we had, and what we were doing, knew that we could easily put something up in space -- and we did do that, including putting up reconnaissance satellite, because it was given a much higher priority because of the Sputnik. I was going back and forth from the West Coast like a yo-yo, and making presentations to the Congress, to the Pentagon at all levels, and so forth.

So it stirred up a fury, so to speak, and a good one. We need to be awakened from time to time, and that really woke us up. But we didn't have a missile gap, as was forecast in the political circles, particularly in the election of 1960. We knew we were ahead of the Soviets, as a matter of fact. We were building the Minuteman solid propellant, and we were ahead of the Soviets on the thermonuclear weapon. We had a Minuteman operational in the inventory, in less than six years after the program started. We were definitely ahead in the solid propellant area, which was much more efficient, from an operational and logistics standpoint.

So we were not comfortable -- don't misunderstand me. We had a burr up you know what. But we felt comfortable in our own knowledge, based on the information that was available, that we were really ahead of the Soviets by a year or so, which isn't much, but we were given very, very good support from political levels, the government, and by the scientific community, and by our own military establishment. So it was one of those dream kind of situations where you've got real support for the program.

And it wasn't just the ICBM; it was also the Navy solid propellant program. I don't believe it's become clear to a lot of people that we were, in fact, ahead of the Soviets at the time that they launched Sputnik. But I thank them for doing so, because it really got us ginned up from the political standpoint, particularly from the space standpoint, because we were getting all the support that we could possibly want in terms of the ICBM and the Polaris Program.

Q: It certainly was a motivating factor.

Schriever: Yes, it was a great motivating factor. I used too many words. That would do it: motivating factor. Put that in red ink. (Laughter)

Q: Okay. One of the motivations that came from Sputnik was the creation of a space agency. Had you expected such an agency to form and that it would be under civilian auspices? What were your thoughts when it was created, and were you involved at all in discussions?

Schriever: Well, yes. There were discussions and there were studies that lasted for several years after Sputnik regarding the creation of an organizational structure for doing it, and there were several different choices that could have been made. The Army felt that their Huntsville (Alabama) facility was the facility that should take over the responsibility for the space business. We [the Air Force] thought that we should be the ones to take over responsibility for space business. The President finally came around and said, "We'll take the NACA organization and the Huntsville organization and put them together and make a NASA," and that was the 1958 act under Eisenhower. Eisenhower was President at that time.

That expanded the NACA (National Advisory Committee for Aeronautics) role, but it put Huntsville in as a beginning facility that moved forward in terms of its capability, but that didn't move the Air Force or what we were doing out of business, but it did put the R&D and the civilian side of the work into the new organization, NASA.

But we, of course, worked very closely with NASA. The first administrator was (T.) Keith Glennan and then came Jim Webb, and Thomas Paine, and I worked very closely with them. Paine came in after I'd retired, but I worked closely with Jim Webb the whole time he was running NASA for about seven or eight years, starting in '61, I think. He was under the (President John F.) Kennedy and (Vice President Lyndon B.) Johnson regime, and I was still on active duty until 1966. So I worked with Jim Webb and not only giving him lip service or things of that kind. We made a lot of people available. We were very much involved in the Mercury Program, the Gemini Program, and in the Apollo Program when George Mueller came over to ask me could he have -- (Gen.) Sam Phillips. Sam Phillips was running the Minuteman Program, he later turned out to be four-star general running the Air Force Systems Command, so you know we were giving him quality.

You hear, here and there, that there's a big feud between NASA and the Air Force, which is not true. We worked together. I knew Jim Webb well. As a matter of fact, when I retired, he helped me get a job or two. So, you know, you get a bad impression sometimes, but I'd like to straighten that one out. Not that we agreed with everything, but, you know, once a decision was made, we worked together and enthusiastically and provided very substantial help. It wasn't only Sam Phillips, but we must have had, in its prime, something like, oh, fifty to seventy-five*Air Force people working full time in the NASA operation, specifically the Apollo Program. [* editor's note: it was closer to 300]

Q: A very important relationship there between Air Force and NASA.

Schriever: Yes, and that same thing was true, we had the job of man-rating the Atlas Program and also the Titan Program.

Q: That's a very important fact in that these were originally designed to be missiles to go out and explode. How did you work to man-rate them?

Schriever: Well, I can't give you a lot of detail, but just one example is the g's -- the forces of gravity -- a machine can take many more g's than a man can, so they had to apply a different burn rate to get the g's down so that the man could tolerate them from a physical standpoint. That was a key thing as far as a man was concerned. Just exactly what had to be done to the missile, it worked, whatever it was. I can't give you the details.

Q: When were the first discussions about using the Atlas for the Mercury Program, do you recall those?

Schriever: No, I really don't recall when they first started, but we were working together with NASA on it right from the very beginning. They were the only boosters we had that could put a man into space. We didn't have anything else. We had the Atlas-period. Then the Titan came along. Of course, going to the Moon was another matter. You had a much larger rocket engine, a complex of engines for that, and those were all developed by NASA. But we had the rocket stands in Muroc Lake in California that were all part of that program as well. We had a very large rocket test capability there and built additional test stands and so forth.

So we were working together with NASA, in addition to just the project. We had people throughout the NASA organization working with them on major programs of that nature. I had General (Osmond J.) Ritland over there for a while, and he was my deputy on the West Coast after that. So we really worked together at the working level, let's put it that way.

Q: In fact, the Air Force was in control of the launch facilities even at Cape Canaveral (Florida), as well, is that correct?

Schriever: Well, yes. NASA had the responsibility for the launch, but it was a team [effort]. The Air Force had the responsibility for carrying out the launch process, and they worked together. I've forgotten who had the "push the button" responsibility, but I presume it was NASA, because it was their program, and we provided the booster to get the astronaut into orbit. We were working together on that, too, in doing the testing and so forth. It was a NASA responsibility, and we pitched in where--well, not only pitched in, but worked together just as a single team. We didn't want to leave somebody up there in orbit, you know, and so forth, or have it crash, or have a failure on the pad. You know what happened as far as when the Shuttle launch went awry (Challenger STS 51-L). It created a tremendous stir. So we had our fingers crossed. We took the risk to bring back the faith of the American people that we could do it.

Q: And, like you said before, those were very successful missions. The boosters worked well, everyone came back well.

Schriever: That's right.

Q: As the Mercury Program was first starting up, in fact, even before anyone had flown on the Atlas, right after Alan Shepard's launch,

President Kennedy made the challenge to send a man to the Moon and return him safely to the Earth by the end of the decade. What did you think when you heard that challenge?

Schriever: Well, I thought it was great. I mean, we had been working on studies of putting a man on the Moon. That started in earnest right after the Sputnik. We were making studies on man to the Moon, and that was before NASA was created. ARPA (Advanced Research Projects Agency) temporarily was given the space mission. I don't know why that was done, frankly, because ARPA wasn't even in existence, and that lasted for about a year. The Air Force was still doing the work, you know, but ARPA was brought into the picture. Then NASA [was created] in '58, which was only about a year after Sputnik. So it took us about a year to create that kind of structure, which included ARPA, incidentally, as well.

But ARPA was given the mission of taking up, you might say, something between basic research and technology; basic research testing, more vigorous technology which the services normally wouldn't pick up. It would be more scientifically oriented. And that worked out extremely well. I had my doubts about that, but ARPA has done a fine job filling a niche in terms of an important part of the overall research, development, test, and evaluation process. If I had to do it over again, I'd probably make a change here and there, but I have no complaints about what this particular arrangement has accomplished, which includes, of course, my Air Force, but it's really working for my country.

Q: You mentioned that the Air Force had been looking into some space programs soon after Sputnik. Do you recall what some of the details of those programs were, the plans that they had been developing?

Schriever: Well, RAND Corporation made a study with respect to space operations and what space satellites could provide, in a study that was completed in 1947, I think. It's available in a report that RAND put out back in 1986 or '87. I have copies of it. Dr. (Louis N.) Ridenour, who was chairman of that particular study, identified every mission that you can think of that would be of value to our national security, which, of course, includes reconnaissance, communications, navigation, weather, and so forth. We were involved in all of those right quick after the Sputnik. Now, they weren't all funded. We had trouble getting some of them funded, and some of them went to dividing who gets what-the Air Force-and there was some confusion, but, nevertheless, the military started on programs, and they weren't all necessarily in the Air Force. Early warning was another which brought radar and infrared sensors into the picture and so on.

So we had a clear definition of what we wanted to do in space to enhance our overall national security posture, and we failed to get all the fiscal support that we felt necessary, and I still feel we could have done better and gotten capability sooner than we have, but, nevertheless, we are the leading power in the world as it pertains to space applications for national security. There's no question about that. It's very important that we maintain that decision. I haven't begun to name all the things that we did. And more commercial activity, particularly in the communications area, has occurred in the global positioning satellite navigation. It's much more than just navigation; it has both a commercial application and a military application from a guidance standpoint. So space has become a [hundred million] dollar industry, and it continues to grow. So it will be important as ground, air, and the oceans, space will probably come out eventually in the first position with respect to commerce and defense.

Q: Looking back at relations between NASA and the Air Force, were there formal agreements made or did everything just kind of flow together? How did that work?

Schriever: Well, there were agreements made in writing. I can't recall them in detail. But there were also, I would say, the major factor of people getting together is working together and having a responsibility and motivation for the job that they're working on. We didn't need any paper that told us what to do with respect to the Apollo Program. We made people available.

I was asked by George Mueller. He had manned space flight under him, and then he had Sam Phillips running the Apollo Program. "Mueller Miller," we called him, he was a TRW man and he brought in and headed the manned space flight activity. Well, we'd worked together for a long time. He came over and asked me whether he could have Sam Phillips to take over the Apollo Program. Well, you know, that's like pulling all my teeth. Here he was running the Minuteman Program, which was the most important missile program that we had, as far as the Air Force was concerned. The Navy would say probably it was Polaris. But they were both important, let's put it that way.

I called Sam in, and I said, "I'll not stand in your way. I think it's a good opportunity for you from a career standpoint. If you want to take the job, I'll make you available, but only on one condition. I'm going to see the Chief of the Air Force and also the Secretary of the Air Force," who at that time was (Eugene M.) Zuckert, I think. Yes, Gene Zuckert. I said, "If I get the green light that they won't forget the people that they send to NASA, forget them when promotion comes around, I'll make Sam available and other people available, but I want to be sure that it's understood that these people are not going to get lost from a promotion standpoint."

And I got sufficient satisfaction that that would be done, and it was done, because Phillips at that time was a brigadier general, and he turned out to be a four-star general. We had a number of-several others promoted while they were doing the work at NASA, so they carried out their word.

So what was done, much of the relationship really related to people getting together and having that team spirit, you know, plus everybody wanted the Apollo Program to be successful. We had a lot of people in the Air Force who could manage big programs and had a very high rating from the industry that they dealt with, so why not make them available? And I think it worked out extremely well. I guess you still have some Air Force people over there.

Q: Absolutely.

Schriever: I don't follow it closely enough. I know (NASA Administrator) Dan Goldin, also came from TRW. I knew him out there quite well, and I see him from time to time.

Q: That's an interesting relationship, too, with TRW, which originally started as Ramo-Wooldridge, working on the ICBM program.

Schriever: Yes.

Q: Can you tell us about how that relationship worked and grew into your work, then, with NASA?

Schriever: Well, it wasn't aimed at that. It was necessary to integrate the major subsystems. If you take a look at the ICBM, starting from the top, you have the atom bomb, the weapon. You have the nose cone, which has to reenter. A lot of people thought we could never reenter without burning up. Then you have the structure. Then you have the propulsion. Then you have the guidance. All of these things had to be put together in one machine, which had never been done before. So naturally we didn't have an industry that was in tune exactly, particularly not any aircraft industry [company]. There were a number of things that other parts of our industry could do better than the aircraft industry, so our decision was that we'd do it on an associate contractor basis. [However,] the integration and the interface of the technology among the various subsystems that I've just given you, that kind of oversight, you might say, engineering-wise, was what Ramo-Wooldridge was doing.

We had no preconceived ideas on how to organize to do this job. It took us about six months of different approaches. As a matter of fact, a lot of people in industry were opposed to the associate contractor approach. They wanted a prime contractor approach. So we had some difficulty in finally making the decision that we'd do it the way I just outlined. Ramo-Wooldridge, had not only a major group of high-quality engineers and scientists, but also the power to draw from universities and so forth. Because this program really got the scientific community ready and able, we really could get the topnotch people in this country on board with Ramo-Wooldridge to do that integration engineering job. That was the background of it, and it was successful.

We had the priority from a political standpoint. We had the authority to make decisions at the working level. We got the good people that were necessary, and we got the job done. It wasn't only the Air Force; the Navy got its job done insofar as the submarine-launched missiles were concerned. These were both major programs, major new challenges, and we both did it. And the management approach was very, very successful and very important to the success of the program.

Q: And you certainly achieved what once was considered impossible. I believe one of the science advisors for one of the Presidents once said that ballistic missiles wouldn't be possible for many, many years down the road.

Schriever: Well, that, unfortunately, was one of the scientists. I knew him well. I can remember having a meeting with him in the Pentagon. He said, "Bennie-" And he was a tremendous man and a wonderful man. "Why don't you take it a step at a time and move forward on a shorter-range missile?" And I said, "Well, we're going to be doing a shorter-range missile, but I don't think we need to do it in sequence." That's what most of the scientists said. I said, "This isn't my decision from a technical and scientific standpoint, but they say we can do it, so I think it's important that we shoot for the Moon, so to speak, and get the range that we need."

So it certainly wasn't unanimous, and particularly with respect to the reentry phase of it. That was the one area that quite a few scientists thought was a very, very, almost impossible thing to do. And we figured out a test program to actually carry out a test of that by taking a nosecone into space and then having enough rocket power to accelerate it so it goes back into space, essentially at the same velocity that our nosecones would be coming back in. We proved it could be done that way, or else we'd have been delayed probably for a couple of years or longer. We called it the 117-L Program, which Lockheed Aircraft did. [It was] a very successful program and it got a lot of people off our back.

Q: Was that the CORONA and Discoverer Program?

Schriever: No, no, it was long before that. That was the first testing that we did with respect to nose cone recovery. The CORONA Program, the Discoverer Program, was really-we didn't make a fast return into the atmosphere. We had slowed it down with a parachute. We recovered it by air snatch in the Hawaii area. So that was a capsule. That would have burned up if we hadn't slowed it down as it entered into the upper part of the atmosphere.

Q: When you were looking at the reentry and bringing the capsule down safely, was there a lot of discussion that involved the heat shielding?

Schriever: Yes, there was a lot of discussion about that early on, and then the heat sink versus the ablative, and the ablative won out only after we'd actually succeeded in making some reentrys with an ablative test program, which the Army did, as a matter of fact. They were there first. We were also doing work on it in the Air Force, but to play it safe, we were going with the heat sink approach on the initial Atlas Program, but we switched over to the ablative, and the ablative worked fine.

Q: Worked fine for the Air Force, for NASA, and got everybody back down safely. We've talked a bit about your interactions between the Air Force and NASA, and we know early in the program you were in command of the Western Development Division, but you moved over to be in command of the ARDC (Air Research and Development Command). How did that change your roles and your responsibilities and the interactions, or did it change it to any extent?

Schriever: Well, it changed it only in the sense that I wasn't running the ICBM and space programs on the West Coast in a daily, detailed manner. It was still part of the command, that ARDC had the responsibility. WDD, or the Ballistic Missile Division, reported to me as commander of ARDC, but I had the whole ARDC, which included the Electronics Division, the Aircraft Division, all the propulsion work, the Armament Division down at Eglin Air Force Base (Florida), test range at Cape Canaveral, and so forth. All of that was part of ARDC, reporting in to Air Force-well, first ARDC, then it became Air Force Systems Command.

I had the responsibility for the procurement of the acquisition phase. Once it started to be bought in a routine manner, then the buying was switched to the Logistics Command, who supported the operational forces. I didn't have anything to do with supporting the operational forces except going through the process of getting something that would work, and once it's working satisfactory, then the Log Command takes over and has the buying responsibility. But the R&D and test phase is over with. That's the way we were operating in the Systems Command.

Q: As you were working with the ICBM program and with NASA, weren't you also involved with some of the Air Force programs such as Dyna-Soar and Manned Orbiting Laboratory (MOL)?

Schriever: Well, I was the director of the Manned Orbiting Laboratory.

Q: Can you tell us about that program and how it evolved and then even how it came to its demise?

Schriever: Well, it had two purposes, one having to do with the ability of man to operate in space over extended periods, but it also had a mission to perform, which I don't know whether it's declassified yet or not, so I won't tell you what it was. But there was an operational mission involved.

A new administration came in and decided that that mission could be carried on adequately by, and more cheaply with, unmanned satellites, and that's what led to the cancellation of the MOL. It was in being for about, oh, four or five years, I guess, maybe longer, but around that time period. So as commander of Systems Command, I was also given the job of being director of the MOL Program, so they reported directly to me and it was a unique arrangement.

Q: Certainly a unique arrangement and a unique program, too.

Schriever: Yes. Well, I think actually that it made sense to determine man's capability. I think that actually NASA had not picked up, at that point in time, [the need] for an orbiting laboratory such as they are doing now, in conjunction jointly with the Russians. I don't know who else is involved in it. Well, you know, after all, we were involved in something brand new, and it wasn't just a plaything that we were dealing with.

Q: As the programs moved forward from the ICBM and then into NASA with Mercury, then into Gemini, you were, I believe, involved on the Gemini Program Planning Board. What did that entail?

Schriever: Well, I can remember one problem. There were several problems, actually, bringing in smoothly the g loading on the individual, on the astronaut. Gemini (Titan), being a two-stage device, had what they called a pogo, uneven application of g loading, which was, of course, a propulsion problem. And I've forgotten now whether it was the first stage or the second stage where the problem was, but we did have a panel or committee set up to take a look at that, and that's the only time that I can remember being involved in that kind of a look-see. But it was important enough that I think I was chairman of it, if my memory's right, but that's the only problem that I can remember that we had with the Titan, which we really were caught out on a limb, so to speak.

We never did get rid of the pogo effect completely, but at least we got it down to a-it was a random kind of thing, so that was a worry. Uneven burning created the g loadings that man couldn't take, so we solved them to the point where we never had a problem with it in flight, but I've still heard some of the astronauts saying they got some fairly substantial jolts.

Q: I know on the Atlas they'd have some weaknesses sometimes in the early launches and would have a tendency to explode occasionally. Were there times when they would look at both the Atlas and the Titan, and when you would look at it, wonder if it was going to be able to accomplish the mission? Or did you think that it would just take enough work and tweaking to make it work?

Schriever: Well, we thought we could make it work or else we wouldn't have done it. You can say that maybe if you'd had something that wasn't as important as showing the American people that we were still in the ball game, you know, then we might not have taken that risk, but there was a certain amount of risk involved. We had five Atlas failures in a row in one instance during the test program, but that was behind us, you know, when we got to the Sputnik thing. This was in the early sixties that the manned space flight started. I've forgotten when the first launch was made.

Q: 1961.

Schriever: '61. Okay. Early sixties. That's right. Well, by that time we had Atlas operational in the inventory up in Cheyenne (Wyoming). Not just one. I mean, we had one there in 1959, I think one or two. You know, you have to look at all factors, and in some instances you take more risks than others. Certainly where life is concerned, that's the highest risk that one takes, because the reaction of the American people is very bad publicity for the guys who did it. "What the hell did you do it for? You should have known better."

Q: We've talked about your interactions with NASA. Were you also involved working with (Wernher) von Braun's group down at Marshall

Space Flight Center?

Schriever: Well, yes. The only times we were very closely working together were on the Thor and the Jupiter. See, they were both intermediate-range missiles, and the Jupiter was being built for shipborne launch from naval vessels, later changed to ground deployment, and they were deployed in Italy. And we deployed the Thor in Great Britain. So they were using the same rocket engine that we were using. The Thor and the Jupiter had the same rocket engine. So we worked together on that fairly close. On most other things, we were not working that closely, because when they got involved in the lunar program, we were not involved in it at all in terms of working with them. I'm sure we had some liaison people there, but they were not really part of his team.

But I knew Wernher quite well. One aspect you have to remember is that I worked very closely with Keith Glennan, worked very closely with Jim Webb. We were working together during the most heated phase of what we do and what they do, and so forth and so on. But from a personal standpoint, we got along very well together, although we didn't always agree. But I had a great respect for him, and I think he respected me, too. But I liked Jim, and he was a little explosive at times, but we had no personal problems. We had disagreements from time to time, but we worked them out.

Q: In general, the space program and the manned space program, in particular, what effect did it have on national security?

Schriever: Well, space overall has had a tremendous impact on national security. We haven't really gotten to the point yet that we understand just how much of a revolution warfighting is going to be, because a major war is very much different than what we're doing now over there in the Balkans (Kosovo / Yugoslavia). I think that it's hard to compare that situation to one where you really have a war. Now, it's a war in the sense of the implements that are used, but the objectives are different. I think that we're in what we call a revolution in military affairs, and it's playing out now.

It's going to be a while yet, I think, before we restructure and rethink some of the ways in which we are going to have to arm ourselves, because we're in the space business now, but there's still that interaction between ground, sea, air, and space, and they have to be integrated, and they are being integrated now, but they weren't really integrated. They did a great job in the Gulf War. That's the first war that I would put in the category of Arnold's brains, and brains are going to play a more and more important role, because the sophistication of precision weapons, the speed of light that relates to information. We talk about information warfare. That's going to have to be integrated in the military actions of hardware and so forth.

So we have a challenge of optimizing our capability in a completely new environment. Space has intruded, you might say, in many ways, and in other ways it can bring about what I consider a spread in our deterrent overall capability. We can deter by--deterrence requires the deterrer to have the credibility that what he has is something that an enemy can't really do anything but, in the end, lose. Then he's deterred. But if he doesn't, for any reason at all, believe that we can do it, then deterrence flies right out the window. So we are in a state of rethinking a lot of things, and I think we've made a lot of progress, but we're still in the phase that is--I mean, we're no longer in the trenches. We talked about bringing people over, bringing ground troops into the Balkans. I'm not going to make any comments on that one way or another, but there are-- we need more time to come out.

When I started flying, when I was at Texas A&M (University), we still had horses pulling French 75s around. Now, mind you, this was 1931 when I graduated there. And look where we are today. So it's an awful lot to swallow, and I think we've done extremely well, but we still have a ways to go.

Q: A ways to go, and we should go carefully.

Schriever: Yes. I mean, ways to go to integrate the ground, air, sea, and space, from a military overall capability. And our first job is to have a military force that deters. The military is there really to prevent wars, and to prevent wars you have to be able to fight them, and they have to believe that the U.S. will win, or a group, a coalition will win, like NATO (North Atlantic Treaty Organization). Using NATO forces has a lot of critics, too. So we're in a very interesting period of history. It may take some years. I'm talking about optimizing what we have. We have a tremendous capability today, but it's not just the military force, it's the political and many other factors that relate to what one does. But the political element has to be an optimized one, and one that will have the least in the way of manpower casualties. And that's what's hard to control in a situation like the Balkans now, and Vietnam earlier and so on. But there are a lot of good brains working it, and it's going to be done, I think.

Q: Talking a little bit about the political involvement in space, when Apollo-Soyuz was first pulled together, this was in 1975 so I know you had retired, but did you have any thoughts on that at the time, of having a joint mission between the Americans and the Soviets?

Schriever: No. I've always felt that cooperative programs (are) one way to eliminate antagonisms and have a better understanding. I think Communism, that threat still exists, it exists in China, and we still have problems.

But I think we have a period here where we do have such overawing capability that we can afford to try to get closer cooperation where you really have a trust, you know, and that this visibility-you know, if you don't trust somebody, you can't really ever make much headway, but the way you trust people is to get to know them, and the only way you really get to know them is work together. I think this period right now is one when, if we can get Russia more westernized so to speak, I think would be a very major step forward in ensuring [more peaceful relations] -- it reduces the emotion that always goes with wars or getting close to a war situation.

Well, let me put it this way. I think cooperation is a good thing, and we ought to try to do it to the maximum extent, but keep our guard up.

Q: Looking back over your career specifically with the ICBM program and then with the involvement with NASA, what was the biggest challenge for you?

Schriever: Well, you know, it's one thing when you are doing it. It's another thing when you review it in retrospect. For example, I never thought that the ICBM program, we were working in the program, and I guess being younger and having access to real topnotch people and so forth, there's no question our greatest challenge was the ICBM program, and creating the management structure that really, I think, was absolutely essential. It wouldn't have been possible if we had not had really major support from the scientific community on that.

As a matter of fact, the committee report that (John) von Neumann headed up (the Teapot Report in February of 1954), was not just a report talking about science and technology and, to get the job done, we now have long-range missile forces and so forth. We got out of that report a portion, that was signed by von Neumann himself, in which he pointed out that we would never be able to get it done unless we changed our management structure so that bureaucracy couldn't stop you at various detailed levels, that you needed a special management approach for the ICBM program. And that's what we spent quite a bit of time on, which I pointed out earlier.

It turned out that we had a unique management approach that's not around anymore, and I think it should be applied to those programs where you really have a major, major breakthrough, from a military standpoint, that you can afford a streamlined management approach. You take a lot of nay-sayers who say no, but can't say yes, and that's a problem that we generally have. Many layers of review, with lots of no-sayers, but they can't say yes. And it exists today. You have to eliminate that. And it existed in the early part of the ICBM program.

Looking back, I think that accomplishing a management approach that is streamlined in the decision-making process, and got top level support, including the President himself, Eisenhower, behind it, probably was the most challenging job I had, but I didn't know it. Because in retrospect, I know a hell of a lot of people were fighting like mad to prevent that management approach to be undertaken, because it broke up a little china here and there, you know, chinaware, not China.

Q: Luckily, you were able to bring that up and meet that challenge and make the program successful.

Schriever: Well, not only that, it proves that management was the key because we hadn't had that kind of [organization before and both the

Army and the Navy did the same thing]. They were bringing things into being, to operational inventory, in five to six years, and that's unheard of, you know, in today's environment. I think time is money, you know. Time is money. And they don't ever measure, hardly ever measure time, except overall they measure what they're paying for what they're getting, but it takes a hell of a lot longer to get it, so you have to add that additional amount of money you spend that's taken up by additional time. Ten to fifteen years it takes to get a new weapon into the inventory, major weapon.

Q: After that amount of time, technology has almost outpaced that system.

Schriever: Well, I don't know about that, but technology is lasting longer now. You have Stealth technology. That's going to last for a long time, but there will be some breakthroughs on that on the other side from a defense standpoint. What they are, I don't know, but now we're talking about defense against ballistic missiles. We thought at one time that here was a weapon that could never be destroyed by the enemy, but I don't have that same feeling now. I think it can be. But I think you can take actions to counter the defenses that might be set up, too. So it's a game of offense, defense, defense, offense, and so forth, so therefore technology continues as long as we have the world that we're living in.

Q: Absolutely, it does. If setting up the management system and making it all work was your greatest challenge, what do you consider as your greatest achievement or success?

Schriever: Oh, I don't know. I guess-well, it's hard to say. I think the greatest success was my opportunity to have assignments that dealt with creating a new force structure as it relates to the Air Force, because I was at Wright Field (Ohio) prior to the war, and at Stanford University (California) when the war started, and came back and had the assignments which I think gave me an opportunity to be involved in what Hap Arnold was talking about, applying technology, new technology, to overhauling, you might say, the Air Force, because we were in the Air Force.

Getting into the long-range missile and space activity, I was a disciple, you might say, of Hap Arnold, and particularly his jet engine, his rocket engine, and the application of nuclear weapons. Thank God they have actually deterred a major war. We haven't fired another nuclear weapon since the one that was dropped on Nagasaki (Japan). That's been quite a few years ago.

Q: Quite a few. That was quite a success, that you were able to bring the program to where it needed to be to (deter war).

Schriever: Well, of course, I'm talking about being involved in maintaining it. I was involved in all of those things, and putting them into what you might call a peacetime environment, although there hasn't been a peacetime as far as regional wars is concerned. So I think our next big challenge is how do we really stop them before they start.

Q: That is going to be quite a challenge.

Schriever: That's deterrence.

Q: Looking back over the involvement between NASA and the Air Force, are there any last thoughts that you have on how that interaction went or how much the ICBM program helped NASA, or any last thoughts on that?

Schriever: Well, I don't really believe the ICBM program helped NASA. I mean, the technology that was involved was important to NASA as well as important to the military. One thing that I commiserated with Keith Glennan and Jim Webb and, since I've retired, with various other administrators of NASA, was that we weren't putting enough money in aviation research, but whether they were pushing it enough or not, the amount of effort on aviation went down.

I worked before the war, when I was at Wright Field as a test pilot and I went to Langley (Research Center, NACA, Hampton, Virginia) quite frequently. I have a high regard for the Langley operation, the Cleveland operation (Lewis Research Center, now Glenn Research Center), and the propulsion area at San Jose (Ames Research Center, California), did a tremendous job.

I think we can't forget aviation. It still needs a lot of additional work. But I don't really know how well the services are working with NASA today, because I don't get at that interface that often. But it seems to me that I don't hear much that--you still have military people working over there, and I think they're all working well together where aviation and space meets. That should be stressed, and I think it is being stressed. I don't have enough knowledge, really, of the details of the operation at the moment, what they're doing in the way of detailed projects and programs, but my message is, keep working together.

Q: Based on your experiences, and this is just speculating, do you see a specific military space agency developing at all, or do you see things just kind of progressing as they are?

Schriever: No. I see a possibility of a Space Force coming into being, from an operational standpoint. I hope it doesn't, because I don't think we need one. But we need an organization that pushes very hard on space and fights the battle here in Washington (D.C.) for budget support and so forth. I think that sometimes I get the feeling that there aren't enough people fighting for that piece of the pie, you know, that's necessary. Look how long it took the Army. I was in the Army Air Corps for more years than I think I was in the Air Force, because we didn't become an Air Force until 1947.

So there's talk about a separate Space Force. I'm talking about logistics and operational responsibility, doing the same function that relates to space that we are doing in the Air Force as it relates to the air. But I personally have always said I'd prefer to have the organizational arrangement stay the way it is, but let's be sure we have the necessary advocacy to push space, because it's that important as far as military operations are concerned.

See, I had four stars for almost six years, I guess, and it's important that we have in the organization-I'm talking Air Force now-it's important that we have someone that is of sufficient rank to be representative of what's necessary in space and who really believes it, you know. We have the Space Command, which is out there at Colorado Springs (Colorado). I think that's very important that that remain a major CINC, or Commander-in-Chief, Space. It's a very important step.

From the standpoint of the Air Force as a service, I think we have to elevate the whole future, the future's part of the-you need a four-star general who's looking in the future, who fights like hell, and that includes space, because that's an area that you're going to need the most advancements in, in terms of operational applications.

I can't name them all, but we need that four-star guy who sits at that decision table and says, "Damn it to hell, I need this and I'll argue with you until the cows come home." You know, you may not win, but you need that advocacy. I don't see it right now. Let me put it this way. I'd like to see it. There's a lot of it; it seems to be more words, and I'd like to see a little more action with the words. Because they're saying the right words, and they're fighting the battle, but I think they can still do better. But as far as changing the organizational structure of NASA, I wouldn't do anything there. Improve internally, you can always do that, and the same thing with any other organization, but overall organization, I think is pretty good. You never can get something that's perfect, you know, in that regard. People aren't perfect either, you know.

Q: Absolutely not.

About the Interviewer
Carol Butler conducted the interview in April 1999 for the NASA Headquarters History Office to record his thoughts on American missile development and other aspects of the space program

An Interview with
James McDivitt

Command Pilot, Gemini 4
Commander, Apollo 9
Manager, Apollo Spacecraft Program
(Apollo 12 - Apollo 16)

On 3 June 1965, James A. McDivitt observed his friend and crewmate Edward H. White II conduct the first American EVA (Extravehicular Activity) during their Gemini 4 mission. Though the Soviets accomplished the first EVA when Alexei A. Leonov walked in space on 18 March 1965, Gemini 4 was a huge achievement for the American space program. McDivitt continued his participation with the American space program as commander of Apollo 9, the first mission to test the Lunar Module (LM). After this mission, McDivitt took the position of NASA Manager of Lunar Landing Operations, in which he directed the final months of effort to land the first man on the Moon with the Apollo 11 mission. After Apollo 11, McDivitt served as Manager of the Apollo Spacecraft Program Office until his retirement in 1972.

***

Q: Who were your role models as a youth and what influenced you to choose a career in aviation?

McDivitt: I didn't have any role models when I was young that had anything to do with aviation. I had never been in an airplane before I joined the (United States) Air Force. I had never been off the ground when I was selected for pilot training. I wouldn't say there weren't any role models but there was no special person that comes to mind. I had read a lot about flying and stories of combat missions from WW (World War) I and II.

Q: Why did you apply for astronaut (training)? Did you feel you were a strong candidate for selection in the second group?

McDivitt: I already had two years of junior college when I joined the Air Force as an aviation cadet. After earning my wings, I had been sent back to school by the Air Force to complete a four year degree. They had sent me to the University of Michigan where I earned my degree in aeronautical engineering. After graduation, I attended the Test Pilot School (TPS) at Edwards AFB (Air Force Base). Edwards was also the site where the Air Force started a new course called Aerospace Research Pilot School (ARPS). ARPS provided the preparation required of pilots to become astronauts. I was selected as one of the first students to attend ARPS because I had done very well academically in the past. After graduating from ARPS, I went back to being a test pilot because of the great opportunities for me to fly a variety of airplanes. These opportunities included being assigned to fly the X-15 and being named as the Air Force's Project Pilot for the F-4. I had no intentions to apply for the second group of astronauts. I was assigned to go over to France for a month to conduct flight tests on some of their airplanes when one day my boss approached me. My boss, the X-15 pilot Bob (Robert Michael) White, had asked me about whether or not I wanted my name submitted on an application for the second group. The results of the astronaut selection were going to come out while I was gone. I told him that I wasn't interested in applying for astronaut (training). When I got back from France, I started feeling that I really owed it to the country to apply. I thought, "After all, they paid for my education." I applied many weeks after the Air Force selection process had officially closed. Bob and I had a long talk about it. He told me that he wouldn't be upset if I went ahead and applied for astronaut (training). I applied with his help, and I got selected. It wasn't the normal route. I'm sure that he had recommended me for the program.

Q: How did the Original Seven treat the Other Nine in the pecking order?

McDivitt: I think they treated us like junior birdmen when we first got there. We were considered their equals after we had been there a few years and had flown a flight of our own.

Q: What were Ed White and Gus Grissom like?

McDivitt: They weren't very similar. I knew Ed very well, and he was probably the best friend I ever had. Ed always wanted to do the job right. He was a risk-taker but not a fool. He was very bright, energetic and conscientious. Let me just describe Ed for you and not Gus. Ed was a very close friend of mine. He and I had gone to the University of Michigan together. He was getting a masters degree in aeronautical engineering while I was getting my undergraduate degree. He didn't have any prior background in aeronautical engineering so we took a lot of the same courses together. We both lived on the same street. Our kids were roughly the same age. We flew a lot together at the University of Michigan. We went to TPS at Edwards AFB together. In fact, we were in the same TPS class. When we graduated, I stayed at Edwards and he went on to Wright-Patterson. But our paths crossed again during the astronaut selection process. He was there with me in Washington when we were both called to be interviewed for astronaut selection. I happened to walk into the same room where Ed was sitting. He said, "I knew you'd be here." I said, "Yeah. I knew you would be here, too." The Air Force did their own astronaut selection process and then submitted a limited number of names to NASA. The Navy and the Marines just submitted everybody without a separate screening process. All of the civilians were submitted in the same manner, too.

Q: How did you and Ed White feel about the full scale EVA (Extravehicular Activity) on Gemini 4 following Alexei Leonov's space walk?

McDivitt: The EVA idea just sort of popped up for Gemini 4, and that was before Leonov's flight. I don't even remember whose idea it was to have an EVA on our mission. Initially, there was nothing even scheduled for EVA of any kind on Gemini 4. But, once it was assigned to the mission, our EVA training was kept in the strictest confidence. We trained in secret because NASA wanted us to be the first ones to do it. We even thought that we would be the first ones. We started preparing for it long before Leonov's flight. Then, Leonov made his flight and blew the whole thing for us. We were surprised when the Russians did it, and disappointed that we weren't going to be the first to do it.

Q: Based on the ground tests conducted were there any concerns about closing the hatch after the EVA?

McDivitt: The distance between the top of the seat and the bottom of the hatch was very limited. I had a very tall sitting height. I recall Stafford and another guy had a very tall sitting height, too. I had to fly up to St. Louis shortly after getting into the astronaut program to demonstrate that I could close the hatch while being fully suited. I demonstrated that successfully. A few weeks later, someone pointed out that the demonstration was done with my back in the vertical position. That wasn't the way it was going to be on the pad. I'd be in a horizontal position and probably stretched out. So I went back to St. Louis and demonstrated it again in the horizontal position. This time we couldn't get the hatch to close. They had to redesign the seat which required reducing the structural height. I managed to fit in the next time we checked it out. But the seat proved to be too high when we got around to do an EVA test, which involved getting back into the spacecraft with the pressure suit inflated. We had to redesign the seat again. However, the hatch wouldn't lock when we went to close it during our Altitude Chamber Test. We decided to finish the rest of the evaluation in our pressure suits since it was near the end of the checklist. They had to bring the test chamber down to sea level altitude before we were able to get out. We showered and cleaned up. I went back out to the Altitude Chamber and got together with a technician to see what was wrong with the hatch. He was working on the little set of gears inside the locking mechanism that was used by the handle to secure the hatch. These gears had a tendency of getting hung up. I was fortunate to work with him to see how the locking mechanism operated because the hatch refused to open on the actual mission.

During the mission, I said, "Ed. I think I can get it to close if we can get it to open." We chatted about that for a minute. We decided to make the attempt even though it was risky. We eventually got the hatch to open. However, the hatch wouldn't lock when we attempted to close it. I was anxious to get him back inside before we crossed over to the nightside of our orbit. I didn't want to end up trying to close the hatch in the dark. Unfortunately, it turned out to be exactly what I ended up doing. I was trying to close the hatch during our nightside orbital pass. I was able to get the hatch to close and to lock after struggling with it for some time. It was awkward because I had to reach around into the corner of the right seat from my left seat, unable to see what I was doing. The geometry of the spacecraft's interior cabin prevented me from having a direct line of sight of what I was troubleshooting. I had to feel my way down into the housing of the locking mechanism while being careful not to puncture the tips of my glove. I had to get my pressurized glove down inside a little groove to touch the gears of the locking mechanism so it would unseize. Anyway, we eventually got the hatch to close and lock. We were supposed to open it up again to throw out our EVA equipment but we had decided against it.

Q: Was there any consideration for a Command Pilot EVA capability in case the pilot performing the EVA became incapacitated?

McDivitt: No. But I had gone through all of the EVA training that Ed did. We worked on it together. They had decided that only one of us would be doing the EVA. We finally made the decision just a few weeks before the flight that Ed would do it. There were no provisions, whatsoever, to rescue him.

Q: Describe liftoff and staging of the Gemini/Titan II.

McDivitt: It was a much higher G ride than the Saturn V or the (Space) Shuttle. I didn't ride on the Atlas but I think the Atlas had a higher G load than the Titan. The Titan's first stage went from about one and one-quarter Gs to six Gs. The second stage went from a few Gs at ignition up to about eight and one-half Gs at burnout. We were zipping along at burnout. It was a fairly high G load. Then, we went right from the high G load to zero G after burnout. The flight of the Saturn V moonrocket was different. I think the S-IC burned out at four and one-half Gs. The S-II burnout occurred at a couple of Gs. We were only at a G at burnout of the S-IVB, which was the last stage of the Saturn V.

Q: How well did the Gemini spacecraft actually fly in orbit compared to the ground simulators?

McDivitt: It was very much like the simulators, though the simulators we had were very rudimentary at that time. They were not motion simulators nor did they have very much in the way of optical simulation. The only optical simulation that we had was for the rendezvous phase.

Q: Your crew attempted to rendezvous with the second stage of the Titan II by sighting the two lights mounted on it. Difficulties were encountered. How much training did the crew receive in orbital mechanics for the rendezvous phase?

McDivitt: I had studied orbital mechanics theory. I was an engineer so I knew how the theory worked. I knew what would happen in theory. But we had no practical training. We originally weren't going to do a rendezvous with the second stage. We were supposed to separate from the second stage and fly in formation with it. The second stage had only two lights mounted on it. We needed at least three lights on that stage to easily see two of them. We needed two lights to get any kind of depth perception. Only in rare occasions would we see two flashing lights simultaneously if we had only two of them on there to begin with. It would have been very difficult to tell if we were closing or opening on the object with only one light in sight. We were supposed to fly in formation with the second stage.

One of things that we were supposed to do after we separated from the Titan was to do a realignment of our platform. The realignment procedure was called "gyrocompassing" the platform. This meant that we had to separate and get stabilized with the upper stage. Our inputs to the spacecraft would no longer be needed once the platform was stabilized.

We were expecting the spacecraft to align with the platform. It turned out that we had forgotten that the fuel valves were closed. This had shut off the propellant to the engines. We went ahead and opened a vent on the second stage to keep it from blowing up. The heat of the Sun would have caused the fuel and oxidizer in that stage to expand if the vent wasn't opened. The whole thing would have blown up. However, the stage just started tumbling around after we opened the vent. The vent acted like a small rocket engine. A rocket engine is nothing but a vent-like structure thrusting out some material. It doesn't have to be a nozzle. It could be just a hole thrusting out some material.

The stage started moving around erratically with respect to our orbit. I managed to put some distance between the spacecraft and the stage while we were gyrocompassing the platform. Then, I tried to fly back over to it. It shouldn't have been too much of a problem during the dayside of our orbit but we went quickly into the nightside. Every time I had thought that the spacecraft was stabilized with the stage, it would either come at us or go away from us. It scared the hell out of me. I chased it around through one-half of an orbit. I was never able to get stabilized with the second stage during that whole nightside pass because of the venting. Finally, we just said, "The hell with it!" We were running out of fuel from our attempts to stabilize with the second stage. We decided to quit.

Q: What do you recall of Ed White's EVA?

McDivitt: We had a very optimistic flight plan. We were supposed to stationkeep with the second stage while Ed was getting ready for the EVA. We were going to do the EVA on the second pass, which was M=2. That notion was totally unrealistic. We realized this after our attempts to either catch up with the second stage or stay the hell out of its way. We just weren't prepared. It was obvious that if we continued with the initial plan that the EVA might not have been accomplished. We delayed the EVA for one revolution. That gave us time to go through the checklist in a rational way. The EVA went along very smoothly despite the hatch problem and the fact that a couple of fittings were left off our suits. We vented out the cabin pressure on schedule. Ed had left the spacecraft on time even though he returned a little late.

Q: Wasn't that because Ed White didn't have a direct communication link with Mission Control during his EVA?

McDivitt: No. He could communicate directly to Mission Control. So could I. It wasn't like on the Moon where we had a separate radio communication system. We were both hooked up to the communication systems onboard the spacecraft in a similar manner. We just ran a longer wire out to him. The tether was an umbilical with electrical wiring that connected him to the radios. It had a big, long hose for oxygen while making sure that he wouldn't get away.

Q: Gemini 4's crew consisted of rookies instead of a veteran astronaut from the Mercury Program like (Wally) Schirra and (Gordon) Cooper. How were the Gemini crews selected?

McDivitt: We actually had some talented guys. The guys that had already been up there talked to us about what it was like. We worked out just fine when it came time for our flight. Our backup crew also consisted of new guys. The crew of Gemini 7, which was our backup crew on 4, worked out just fine for their flight, too.

Q: Your crew wanted to name the Gemini spacecraft "American Eagle" but NASA management wouldn't allow it. Why weren't nicknames allowed for spacecraft?

McDivitt: George Mueller insisted that we fly by the mission number. We had to fly by the name "Gemini 4." He had stopped us from calling the spacecraft by any other name. But, I eventually convinced him into letting us call both spacecraft used in the Apollo Program by nicknames. In the beginning, he was insistent that we would be able to call both spacecraft by the name "Apollo 9." I said, "We just had some nicknames that we were using." He said, "No. You cannot do that." I said, "Alright. I will agree not to do that if you agree to come to our next simulation." He said, "Well, I don't know." I eventually forced him into agreeing to come to our next simulation. It was going to be a rendezvous simulation where we would use the name "Apollo 9" for both spacecraft.

Q: That would be confusing.

McDivitt: It sure as hell was. The ground would call the name "Apollo 9" to give us instructions. We said, "Roger. This is Apollo 9." They said, "No, no! Not that Apollo 9. The other Apollo 9!" It was biggest god-damn fiasco that the world had ever seen. Mueller finally had to relent. I made him start while he made me stop.

Q: Your crew was the first to wear the national emblem on the pressure suit. Whose idea was it to have the American flag on the G4C (Gemini space suit made by David Clark Company, Inc.) pressure suit?

McDivitt: That was our idea. Ed's and mine. We did everything together.

Q: What about a mission patch for Gemini 4? The mission patch first appeared on Gemini 5. They were referred to as "Cooper patches."

McDivitt: We wore the American flag.

Q: Would you consider that as your mission patch?

McDivitt: That was sort of our patch. We decided that we ought to have the flag on our pressure suits. We were the first ones to add anything to the suit. We added the flag. We didn't add a patch.

Q: Were you ever considering adding a mission patch?

McDivitt: No. We had never even thought about it. Unfortunately, some other people have created a patch that they said was ours. But, we never had a patch. The flag was closest thing that we had to a mission patch.

Q: The same could be said about the patches for the Mercury flights and the Gemini 3 mission.

McDivitt: Right. I don't know if Gemini 3 had one. They called their spacecraft "The Molly Brown." But I know damn well that we didn't have a patch. We looked at the American flag as sort of our patch. We were the first ones to have the American flag on our suits.

Q: It has been worn on the pressure suit ever since your flight.

McDivitt: Right. We were the first ones to wear the American flag. It has appeared on every mission since our flight on Gemini 4.

Q: Your (Apollo 9) crew worked on finding a way to free the jammed ATDA (Augmented Target Docking Adapter) shroud on the Gemini 9 mission. There was even a look at having Cernan perform an EVA to cut the straps that held the shroud on the ATDA. Did you feel Gene Cernan could have safely cut those straps on the ATDA?

McDivitt: No. I don't think he could have succeeded. My Apollo 9 crew was out in California conducting some spacecraft tests when we received an urgent call. We were asked to go over to Douglas to see what we could do with a pair of cast scissors. Cast scissors are used to cut off casts for setting broken bones. This was the only tool that the Gemini 9 crew had onboard. We raced over to the nearest hospital and acquired a pair of cast scissors, which I failed to return.

Q: Do you still have them?

McDivitt: I still have them. I saw them a few years ago. Anyway, we raced over to Douglas. They had the mock-up put together to what they thought it looked like. They were not able to see for themselves as to what shape it was actually in, so they had to rely on the crew's description of its current condition. We went around it and looked inside during our assessment. We were trying to figure out how the hell we would cut the springs and the big metal band that was wrapped around it with a pair of cast scissors without getting the suit punctured. Finally, we concluded that it was a bad idea. We recommended against it.

Q: You had been transferred over to the Apollo Program instead of being assigned to command another Gemini mission. Did you have any reservations of being transferred over to the Apollo Program Office?

McDivitt: No. Not really. I commanded Gemini 4 and capcommed Gemini 5 before going over to Apollo. My transfer was regarded in the same light as to Gus being assigned to look after the developments of the Gemini spacecraft. I was assigned to look after all of the engineering developments in the Apollo Program.

Q: Describe your work with the Apollo 1 crew on the Block I CSM.

McDivitt: I was the commander of the backup crew. We had worked together on it for a long time. Rusty (Russell Schweickart) and Dave (David Scott) were the other members of the backup crew. We had worked with Roger, Gus and Ed from the very beginning. I think that we had worked on it even before Ed was assigned to the prime crew. Ed was the commander of the backup crew for Gemini 7.

Q: What were your feelings on the Block I design before the Apollo 1 fire?

McDivitt: The Block I had a lot of problems but that was not unusual for a new spacecraft. It was a very complicated piece of engineering. We did have a lot of problems with it during testing. The six of us had spent an ungodly number of hours working on the spacecraft. We had participated in all of the tests for the Block I. We had written the spacecraft's test and operating procedures. We had spent a lot of time on its development.

Q: Did you have an equivalent of a NATOPS (Naval Air Training Operation Procedures Standardization) to operate and fly a CSM (Command and Service Module)?

McDivitt: Yeah. We had these great big procedure cards that were velcroed on to the instrument panels for different phases of the flight. We had made these cards up ourselves. The backup and the prime crews did that on their own up until the time of Gemini 4. We realized that we needed more support starting with the Gemini 4 mission. We had recruited some of the guys from our Flight Crew Operations Directorate to help us out. Deke (Donald Slayton) was in charge of this big organization. It was made up by personnel from the Astronaut Office, the simulation people and the aircraft people. It also included a group of engineers that later became known as the Flight Crew Support Division. Those guys had helped us out since we hadn't specifically assigned anybody to this role. They had assigned us one guy. I think that he may have had one or two assistants. But, generally, it was just this one guy who had carried the major part of the workload. We had the prime crew, the backup crew and this one guy with his assistants to get the job done. By the time of Apollo, the workload had become so astronomical that our crew had created another group called the support crew. We now had a prime crew, a back-up crew and a support crew to handle the increased workload. All of these crews consisted of astronauts. The support crew knew that they were not going to fly the mission even if both the prime crew and the backup crew were wiped out. Another crew would be chosen to fly it. The support crew was vital to what we were doing. The support crew positions were filled with the less-experienced guys but they were invaluable by the time we had reached the flight of Apollo 9. We now had to manage two spacecraft starting with that flight. The support crew had helped us test both spacecraft. In fact, the LM (Lunar Module) was brand new. It was in a state similar to what the CSM had been in its early days of development. They had worked out a lot of the problems in the simulator. They had also developed the procedures for the checklists. They were a big help. Moreover, the flight crew support group was available if we needed help in accomplishing other tasks.

Q: Based on Deke's book, you were originally assigned to the Apollo 2 mission using AS-205/208. It was being planned as a "D" mission, essentially a test of the LM in Earth orbit. How did you feel about your assignment to that mission?

McDivitt: We had never called it Apollo 2. I was never assigned to fly on AS-205. That flight was going to use the Saturn IB. I was originally assigned to be on the backup crew of Apollo 1. I had been assigned to that position for some time.

Q: Wasn't Wally Schirra's crew originally assigned as the backup crew for Apollo 1?

McDivitt: No. He was not. Wally and his crew were assigned to fly on AS-205. I don't even remember who was on their backup crew. They were doing a pressure test on the Block I Command & Service Module (CSM) for AS-205 when it blew up. There wasn't a replacement spacecraft for that mission. Apollo 1 was dragging during this time. Then, one day I was called into Deke's office and told that I would command the first Block II mission with a man-rated LM. I had switched over from the Apollo 1 backup crew to that assignment. This switch resulted in Wally's crew, who were out of work, so to speak, in being assigned as the backup crew of Apollo 1. They were the backup crew for only a month before the fire broke out. Gus had asked me to continue to work on Apollo 1 because my crew had completed so much of the job already. I can recall on a number of occasions that our crew was sent somewhere to go through the procedures. We had to check out on what they were going to do on the mission. We wanted to make sure that it was consistent with Gus' philosophy. Wally didn't know anything about his philosophy since he was a latecomer to the mission.

Q: Do you feel that we were going too fast and that our pace might have been the cause for the Apollo 204 fire?

McDivitt: No. The President said that we needed to get to the Moon before 1970. And we did it. I don't think that we were going too fast. In fact, we would have never made it if we had gone any slower.

Q: According to Deke's book, your crew would have originally flown on Apollo 8 followed by (Frank) Borman's crew on Apollo 9. NASA switched the crews since LM-3 wasn't ready. Was your crew offered the lunar orbital mission or "C" prime mission?

McDivitt: Not really. I know that Deke had said otherwise. He and I were the only guys there. And that wasn't what he told me. He said that he wanted me to continue to fly the "D" mission with the LM.

Q: Were there any problems for the crew to switch from CSM 103 to CSM 104?

McDivitt: No. By that time we had become good spacecraft testers. We had tested the first Block I, the first Block II and the second Block II. We had tested the first LM, LM-2 and LM-3. We had quite a bit of spacecraft testing under our belts.

Q: Was there a lot of pressure to deliver positive results on Apollo 9 since it was the first full up test of the CSM, LM and the EVA pressure suit?

McDivitt: There was a lot of pressure on all of the Apollo flights. We had completed the "C","D","E,"and "F" missions. We had flown "C" successfully. Apollo 8 wasn't a scheduled mission. It was an ad hoc or fill in mission. We had combined the "D" and "E" missions on Apollo 9. We still had to do an "F" mission. We knew if any one of these missions failed that it would have to be reflown. We flew the "F" mission and it worked out fine. Our next flight was the "G" mission which was the lunar landing. There was a lot of pressure on everybody to make sure that mission succeeded.

Q: What were the liftoff and staging like for a Saturn V?

McDivitt: The liftoff of the Saturn V was a lot more impressive than a Titan because of an increase in the acoustic vibration. Just the noise that came beating around the CSM was readily apparent. Being part of the crew, I was the closest person to those engines even though they were several hundred feet below me. I can imagine that the noise was tremendous even to a spectator who was standing three or four miles away. It was hard to imagine what it was like being right there. Those engines were so powerful that the whole thing was vibrating at lift off. The vibrations were caused by both the mechanical operation of those engines and the acoustics generated by the noise. It was a very loud, shaky and bouncy ride from the time of liftoff to the point that the booster went supersonic. The supersonic regime was where we had left all the noise behind the booster. We were traveling faster than the speed of sound at that point. Consequently, the sound below the rocket couldn't reach us since we were traveling supersonic. Most of the vibration stopped and the G levels eased up as soon as we reached supersonic speed. There was a big jolt when we staged. The whole Saturn V was like a big spring that was compressed before those engines shutdown. We had shutdown all of our engines at once. It was literally eight million pounds of thrust that had been cutoff at that moment. The whole rocket responded like a compressed spring that had been released. We were thrown forward and backwards a few times in our couches. We became familiar with the instrument panels from a finite distance.

Q: Recount the docking and extraction of the LM from the S-IVB?

McDivitt: Dave Scott was flying the maneuver when we separated and started our turn around. I was sitting in the middle seat, a position that I couldn't see out of so well. He said, "Gee. You know something's wrong. I can't fly this thing. Something's wrong. I can't fly it!" I said, "Are you sure?" He said, "Yeah. It's not going to where I want it to go."

So I checked over the instrument panel and discovered that the shock of the explosive bolts separating the CSM from the SLA (Spacecraft Lunar Module Adapter) had closed a whole bunch of valves in our Reaction Control Systems (RCS). I just went over and opened them up again. Dave did a good job after that problem was solved.

Q: What kind of mission would have Apollo 9 been if the S-IVB had a premature engine shutdown on the J-2 during the first burn?

McDivitt: There were different kinds of abort modes. We could have separated and flown into orbit on the SPS (Service Propulsion System) if that had happened. The key factor was that we had to be high enough in altitude to make the attempt.

Q: What about an SPS failure after initial LM maneuvers to set up for the rendezvous and docking phase? I would think that the LM would have to become the active element in the maneuver.

McDivitt: In our mission the LM performed the docking maneuver anyway. We were the only mission that flew the LM in this manner. It was too tough to do. I did it but it just wasn't worth it. We were looking out of the LM's top window while the whole flight control system was set up for use in the forward facing direction. We had to do a ninety degree transformation in our heads. Yaw on the yaw controller now became the roll. Roll on the roll controller now became the yaw. It was screwy that we had to do all of these conversions in our heads. I think that Dave forgot to turn on the light for the docking target on the CSM. The circuit breaker was out when I got back over to the CSM. But, I was having a bitch of a time seeing it. I don't know whether it was the popped circuit breaker or just the CSM's shiny surface that had caused me to have so much difficulty in locating the target. It was very difficult for me to see that little target from my overhead window.

Q: Could the CSM have easily changed orbits to rendezvous and dock with the LM if the LM's Descent Propulsion System (DPS) or Ascent Propulsion System (APS) failed to operate?

McDivitt: Yeah. We had a lot of backup procedures.

Q: How would the crew transfer be accomplished if the CSM and LM were unable to physically dock?

McDivitt: We would just do an EVA.

Q: Was this rehearsed at all on the ground?

McDivitt: We had practiced the procedure in the water tanks.

Q: How would the spacecrafts be set up for the EVA transfer?

McDivitt: We would have to put the LM in a passive position and turn off all of the rocket engines. We didn't want to be climbing around out there with the rocket engines on. The CSM would have to come over to retrieve the LM crew. We had sat down before every flight and had very carefully thought through all of these contingencies.

We would have a mission rule out that we had all agreed to beforehand. We didn't ad hoc this stuff while we were doing it. Every flight had a set of mission rules. We would do this contingency if this system failed. We had thought it out very carefully in the coolness of a conference room rather than the heat of battle. We just didn't think through the rules once, and say, "Okay. That's it." We would write up the mission rules as to what the guys who were working on them had seen fit. Then, it was distributed to the astronauts, the flight controllers, the program office and everybody else to get their feedback. Some people would say, "Oh, God. That's stupid. We shouldn't do that. We should do this," and we would redo them if their point was valid. It was a very long and thoughtful process that went on before the mission rules were finalized.

Q: Were you ever approached for command of a lunar landing mission?

McDivitt: Yes. I could have commanded Apollo 13 if I had wanted it.

Q: Why did you turn it down?

McDivitt: I wanted to become the Program Manager. Landing on the Moon to be fourth or fifth wasn't a big deal to me. I mean, "So what!"

Q: It still would have been a fantastic experience.

McDivitt: Oh, yeah. It would have been an experience but so was running the program. In retrospect, I think that I learned a hell of a lot more by running the Apollo Spacecraft Program than I would have by making a lunar landing. It was more in line of what I wanted to do.

Q: You were offered (the opportunity) to run the Astronaut Office and U.S. Air Force's Manned Orbiting Laboratory (MOL) Program. Why did you turn down these offers?

McDivitt: I wasn't interested in running the Astronaut Office. I'd run it part-time when the other guys were gone. That was no big deal. There were offers made to me in running other big programs. For example, General (Samuel C.) Phillips had asked me to run the MOL Program. I flew out to the MOL Program Office in California and talked with the people out there. I had on a couple of other occasions visited and checked out a program's operation when it wasn't familiar to me. But, based on my visit, I concluded that the MOL Program was not going to survive. I told Gen. Phillips that I didn't want to run it because of the conclusion I had reached.

Q: And you were right.

McDivitt: And I was right. I was also offered the opportunity to fly in the DynaSoar Program which I turned down for exactly the same reason.

Q: What attracted you to the position of Manager of the Apollo Spacecraft Program Office?

McDivitt: I liked the challenges. It was becoming obvious that just the rotation of the crews was going to prevent me from being the first man to walk on the Moon. I wanted to do other things in my life than just be a test pilot and an astronaut. I wouldn't trade it for anything in the world. But, test pilots leave the position either by killing themselves, failing a physical or quitting. The first two choices didn't seem like good alternatives to me so I decided to leave. I would resign or leave when the time was appropriate. I also wanted to run a big program, preferably one on the government level. I had spent a lot of my time while I was an astronaut in the management side of the programs. Even during Gemini, I had spent a lot of time attending the program meetings that didn't have any- thing to do with flying the spacecraft. I had assisted the Apollo Office on numerous occasions in solving a problem on the spacecraft. There were other areas that I was involved in that didn't have anything to do with being an astronaut. I began to really enjoy it after spending a lot of time doing it. I wanted to run a big program. I can remember when George (M.) Low called me and asked if I would like to be his successor. I thought it over for about ten seconds before I said, "Yes!" I also wanted to retire and become a civilian. I wanted to do other things in my life besides the military. I was already a very young general. I had only eighteen and one-half years of commissioned service when I became a general. I still had seventeen more years to go as a general. I made the rank very early in my Air Force career and probably would have had a great time at it unless I tripped by stepping on my tie.

But, I also wanted to do other things, too. And I went out and did them.

Q: What factors influenced you to give the green light when some of the Apollo flights degraded into potentially dangerous situations?

McDivitt: The key thing with respect to all of these missions was that I always looked at it as an investment in risk. There was no investment in any risk if the mission was still on the ground. If there was a problem, a decision based on zero risk investment was made. The decision could only go two ways. It would have been easier to call a delay but by delay- ing it for another month there would have been an increase in risk at the time of launch. The spacecraft's RCS had a tendency to deteriorate from its exposure to its own hypergolic fuels after a period of time. But, a very critical part of the mission has passed once the mission was successfully launched. A big investment in risk was made at the time of the launch. There was no turning back once the decision to launch was made. After the launch, I would be a lot more aggressive in going forward with the mission. The crew had risked their lives to get up there. This same rea- soning was applied to the problems that occurred on Apollo 12, 13, 14 and 16. I had never thought that I took any unreasonable risks with those missions. We had thought through the mission rules very carefully before we reached a decision. We would go ahead with the mission as long as we had a backup system. I tried not to do anything irrational. I don't think I ever did. We were very aggressive in trying to accomplish the goal, which was to land on the Moon. The purpose of the flight wasn't just to go up and immediately come down.

Q: Do you feel the crew of Apollo 15 was reprimanded too harshly for the stamp scandal?

McDivitt: No. They probably weren't reprimanded harshly enough.

Q: I assume for the last two lunar flights that the policy was strictly enforced. How did they police it?

McDivitt: We had a very strict policy from the very beginning of the lunar landing missions. They had just violated the policy. It was Deke's job to provide me a list of all the stuff that they were flying onboard. He didn't do that.

Q: They must have become strict on the policy for the crews after Apollo 15.

McDivitt: They were supposed to be strict from the very beginning of these flights.

Q: How did you feel about Joe Engle being replaced by Jack Schmitt as Lunar Module Pilot for Apollo 17? Any talk about switching the entire crew of Apollo 17 with 18?

McDivitt: There wasn't any Apollo 18 mission. The program sort of pooped out before Apollo 18 went into the planning stage. I wasn't going to be there by the time Apollo 17 flew so I was neutral. I'll tell you one of the things that I did feel though was that everybody we had sent up there was qualified to do the job. Engle or anybody else could have done the job as well as Jack. There wasn't a hell of a lot of geology done on the Moon other than picking up rocks and taking pictures. The guys that went up there were well trained in geology. Hell, Jack was a geologist, and he hadn't been out looking at rocks anymore so than all of the rest of the astronauts. It was a political issue to take a scientist there.

Q: Were they were considering a landing on the farside of the Moon?

McDivitt: Well, there was talk about a farside mission. But, there was no way that we could have covered it without putting up a communications satellite. We didn't have any communications satellite to put up there. It would have taken a long time. Yes, there was talk about going over and landing there, too. The solution wasn't as simple as just sticking a satellite in lunar orbit along the farside of the Moon. It wouldn't have fulfilled the requirements for a communication link. That setup wouldn't have worked in Earth orbit, either. We would have probably needed between two to four communications satellites in almost a lunar synchronous orbit to insure that we would always have one covering the farside. This would have allowed a constant communication link for a farside mission. It wasn't a very simple thing to do.

Q: Do you feel NASA management did everything in its power to save the Apollo 18 through 20 missions from cancellation?

McDivitt: I think NASA management wasn't very keen on going forward with those missions. But that was only one part of NASA management because I was in another part. I thought that we should have continued in going to the Moon. We had made that humongous investment in going to the Moon. We had the opportunity to go there three more times. That would have been a fifty percent increase in what we had already accomplished. We had a great capability in those spacecraft for the last three missions. We could have placed everything we did on Apollo 11, 12, 13 and 14 into one J series mission. I think that we missed a golden opportunity. They had this great concern about killing somebody up there but that wasn't my position.

Q: How would you summarize your experience with our space program?

McDivitt: I did my job. We had accomplished what we set out to do. It was a great personal satisfaction for me. I think, in itself, it helped the country and a lot of people.

About the Interviewer
Don Pealer earned a B.S. in Aerospace Engineering from Boston University in 1988 before becoming a Naval aviator.

"Remembering the Space Race" with Walter Cronkite

In the 1960s and early 1970s, there was an extraordinary interplay between television journalism and our national space program to reach the Moon. The space program provided a well-organized succession of spectaculars, and television new provided the crucial means to bring this historic epoch to a global audience. On November 15, 2000, the National Air & Space Museum at the Smithsonian Institution hosted a panel of network news executives and NASA officials to discuss how NASA and the networks collaborated to present the space race.

Dr. Martin Collins
The space race was unique - in what we and the Soviet Union accomplished, and, at least on the American side, how we followed the story. Through television, more than in magazines or newspapers, we were participants, right there at the launch pad, in mission control and in capsules with the astronauts. We were surrogate explorers. This mass participation in the exploration of a harsh and dangerous frontier was a new thing. We had to learn about new science and technol-

ogy, a new language of pilots and capcoms and, occasionally, reflect on why we undertook this exploration and what it meant.

The space race was suited to television. It featured astronaut heroes, risk, potential of catastrophe, and profound firsts -- first man in space, first space walk, first trip around the moon, first human steps on another heavenly body. All pursued against a backdrop of a Cold War, a global contest with the Soviet Union. The Cold War was the rationale for mobilizing resources on a national scale. Hundreds of thousands of government industries and universities contributed their efforts to send a few people into space. The political stakes intensified the drama.

Short Biography of Panel Participants

Walter Cronkite: Famed CBS news anchor.

Jim Kitchell: Jim joined NBC in 1953 and spent 28 years there before helping to create CNN. Among his notable accomplishments at NBC was as director of the Huntley-Brinkley Report.

Robert Wussler: Having worked his way up from the CBS mailroom, he eventually became an executive producer and then president of the news division of CBS . He later moved on to CNN and Turner Broadcasting and was a key player in the development of Superstation TBS and the introduction of TNN. In 1986, he was the general manager of the Goodwill Games in Moscow. He was also the Chairman of the Board of the National Academy of the Television Arts and Sciences for two decades.

Joel Banow: Mr. Banow is an award-winning director for 38 years in broadcast and theater. As director of CBS news special events, he was the first new director to win a Director's Guild Award for outstanding television direction for his work on Apollo 11.

Julian Scheer: Formerly a newspaper reporter, Julian served as NASA's first Public Affairs Director, beginning in 1962.

Dr. Martin Collins: Curator, Smithsonian National Air & Space Museum

The Soviet Sputnik started the race in October 1957. This stunning first leap into space pushed the United States to develop three space programs: one, the military space program; another, an intelligence program to spy on the Soviet Union; and the third created the National Aeronautics and Space Administration which led us to the moon. It was NASA's policy of open access, established gradually over time, to the media that helped to feed an American fascination with science and technology, with exploration, with heroes who had the right stuff and turned the tensions of the Cold War in a more peaceful direction.

NASA and television brought us along on our greatest feat of exploration.

Impressions of the First Days of the Space Effort
Life As A Reporter Trying to Cover This New Beat

James Kitchell: Having spent a little bit of time before NASA at the Cape, it's interesting that at the time that Sputnik was launched and, of course, the failed initial attempt at Vanguard, the space program was really tripartide in that it was military in it's origins. All three military services were running their own programs. Vanguard, although it was a scientific adventure, was being done by the Navy. The Army, with Wernher von Braun, at Huntsville in Alabama, was running their programs of Jupiter and Redstone and so forth. And the Air Force was developing intercontinental ballistic missiles. So there was an awful lot of competition that was going on.

Coverage of the story was somewhat difficult. It was a little bit of a cat and mouse game. I remember one incident before we had any agreements with the military officials where we were trying to cover a launch with film cameras, and the provo marshal sent a helicopter over the top of us and stirred up the sand around our cameras so that we couldn't take pictures. And I had a few incidences with my next-door neighbor, with some property that I had rented down there for the purpose of covering the events. He said, you're spying on the government and tried to keep us from going into the house. There were a lot of things like that that were going on.

But ultimately, in about 1957, early 1957/ mid 1957, there was a private agreement among seven media representatives to allow the press onto the Cape -- and it was a turn around.

Vanguard had been a disaster. The Navy had tried to set up a public relations event of their own but had set up a press site off the Cape and, when it blew up on the launch pad, nobody knew what had happened. All they saw was a big orange ball. So it took some time and, over a couple of years, there were agreements to allow the press onto the Cape. It had become an interesting story to the public, and we moved forward in what we were doing, kind of haltingly. And then the competition started.

Walter Cronkite: Jim will remember, as well, that we had no information except that we might possibly obtain from informants, who were among the engineering staffs that went out to the Cape Canaveral. We were locked up, practically, in the little town of Cocoa Beach. And being locked in the little town of Cocoa Beach was very similar to being locked in Las Vegas for a month. (laughter)

There were about five bars, and we moved from one bar to another. That was about all we had to do while we waited for a shot, which we didn't even know what it was. We weren't necessarily sure what kind of a boost vehicle, what kind of a rocket they were going to fire. But we hung around these bars, and when the engineers all disappeared from the bars one evening, we decided that there must be a shot about to take place. This was our source of our information. And while Jim, working for NBC, hired a house, where his people could sit and look over toward Cape Canaveral, and, perhaps get a picture at long distance through the tight lens, of a gadfrey with a rocket in it.

My boss, Bob Wussler, didn't bother to hire a house at all. He saved CBS a lot of money, and, therefore, got promoted for it. (laughter) When the engineers disappeared, we went out to a miserably, cold, hard naturally granite breakwater, that was the breakwater to the entrance to the Port Canaveral canal. We, with other reporters from other equally cheap services, (laughter) climbed out on these rocks -- this was the most dangerous part of the whole space program -- we climbed out on these rocks. We perched on these miserable rocks during cold nights, and all we had to look at was a bright light over there at Cape Canaveral, a couple of miles away, which we assumed was a rocket gantry, and from which that bright light, a rocket would rise sometime, perhaps, during the night. It was very difficult for us, but we could, at least, turn our backs every once in a while. But our cameramen, our poor cameramen couldn't take the chance of that rocket going off and their not getting it. So they stood there in these cold and miserable circumstances perched on these rocks, looking through their eyepiece, their finder, for hours, waiting for that thing to go up. And most nights, it didn't go up.

We sat there throughout a lot of the evening until a very kindly inn keeper down there, a fellow named Henry Landsworth, who ran the Satellite, one of the hotels, he realized how we were suffering out there. As a matter of fact, he had been sending us food and sustenance of various nature. (laughter) He realized that he could help us a great deal. When the engineers came back from the Cape and came in to drink again, he knew that the mission had been scrubbed. He'd sent a messenger out to us. He probably saved more lives than were saved by the Red Cross in World War II. (laughter)

Robert Wussler: There was an interesting story about Henry Landsworth. I wasn't going to use his name before, but since Walter revealed his existence. Henry was the manager of the one existing decent motel in Cocoa Beach at that time. I remember coming up one night after driving from Miami and getting to the motel about 10:00 at night. The usual question was, who's in town. You got this answer, well, general so-and-so, and so-and-so from General Dynamics, and so forth. Oh, that's interesting, Henry. Because that meant that there was an Atlas launch coming up.

And I said, well, Henry, I've had a tough day; I'm going to go to bed. Why don't you give me a call about 7:00 in the morning? And he looked behind the counter and said, might be a little late (laughter). Oh, we're that close. So then I said, well, why don't you call me about 1:00 in the morning? A little early. (laughter) Why don't you get up about 5:00, and go out on the beach for a walk?

Walter Cronkite: Later on, Henry Landsworth poisoned half of Cocoa Beach. When John Glenn finally got off on his first trip, his first orbital trip, Henry had arranged to have, at that time, what he thought was the world's largest cake baked for the occasion. It took a truck to cart it out to the back end of his hotel where he kept it under a tent. The Glenn launch was delayed, I think, seven times over a period of the almost two months, and, in the hot Florida sun, this cake rebaked under the tent. (laughter)

He still managed to keep it a secret until Glenn came back and, of course, came back into Cocoa Beach, and he unveiled this cake. Well, he fed it to the population of Cocoa Beach. Several members of the community of the Cocoa Beach, most of them, didn't appear on the streets for two days (laughter).

Henry was a wonderful character. He was a boy survivor of the death camps in Germany. His family was all killed except for Henry and a twin sister. They were 14-years old or 15-years old when the war ended, and he came to this country and made a great success of himself which included, in the last 10 or 15 years, a partnership with John Glenn in building hotels on the perimeter of Disney World. They've done very well.

Robert Wussler: Henry Landsworth's nickname, at the time, due to the fact that there were only a limited number of rooms in Cocoa Beach, was either "Henry roll-away" or "Henry double-up". (laughter)

Julian Scheer: That cake was baked in October, and we ate it in February the next year, February the 20th of the next year. It was baked about the 15th of the October. They had two delays; they finally launched. But, you know, going back to the genesis --this is all pre-NASA. I was a print reporter at that time, and we never thought television guys knew what they were doing anyway. We didn't worry about Kitchell. He was always in the bar.

James Kitchell: That's where we got the information.

Julian Scheer: Well, you have to go back to the end of World War II to look at, really, the genesis of the space program. There was something called Operation Paper Clip, and the Russians and the United States both rushed to Penamunde to where the Germans were launching the V-2 rockets. We snatched as many of the German scientists as we could, and the Russians got the rest. We got von Braun and his group, and they came to the United States. We gave them immediate citizenship and put them to work in Huntsville, Alabama on something called a Hermes rocket, the first rocket, which later became the Redstone.

What you really had when Sputnik went up in 1957 was a competition between old friends from Germany: some in the Soviet Union, some in the United States. And that was, really, the genesis of the U.S. space program.

And, as Martin mentioned earlier, we had really three space programs at the time. The Naval Research Lab, here in Washington, had the Vanguard program. The Air Force was really not much interested; they were still interested in airplanes, air-breathing instruments, and they thought there was no role for the missiles. The Army was interested in artillery so they developed a missile program at Huntsville, and out of that missile program came von Braun and the Redstone, later the Saturn, and everything that we used in the Apollo program.

The shocking event for the American people was the October 18, 1957 launch of Sputnik. That really set the tone for everything we did in space after that.

The Russians had the first satellite, the first dog in space, which was the second flight, the first man in space, the first woman in space, the first space walk, the first impact on the moon, the first soft landing on the moon, the first impact on Venus. So the whole nature of the space program after that, after Sputnik, was really a race with the Soviet Union up until Apollo 8, when the Russians decided that they were not going to be first. Essentially, the program ended at that point in time, and we went ahead with Apollo 9, 10, and, eventually, 11.

THE 1961 FLIGHT OF ALAN SHEPARD
The CBS studio was in the back of a station wagon

Walter Cronkite: ...That poor NBC guy was standing out in the field, kicking away the rattlesnakes and the water moccasins, and fighting away the mosquitoes and Wussler had this great idea of talking the NASA people into letting us come out with that vehicle [a station wagon] in which he crammed me in the back, and said, we don't want any noise coming in so keep the windows up. It was only 140-degrees. I think he had a favorite announcer he wanted to get in there instead of me. (laughter)

James Kitchell: We're not going to start with the one-ups-manship here.

Walter Cronkite: You know, it is true, through. That is a reminder of what we had to deal with in the early days. In the earliest days, as I was saying a moment ago, we didn't get on the Cape at all except with very special privileges, and that was for an hour under guard or we took a quick look. We never were able to be there for a launch of any kind under the military regime.

But when these fellows at NASA took over, they quickly realized that if the space program was to get -- the civilian space program -- was going to get the kind of support that it was going to have to have, the billions of dollars through Congress, it had to have the public's approval, and, therefore, public relations became a part of the game. And Julian became part of the game. Then we were taken out under rather severe guard, still, I think the military insisted on that. We were taken to a location where we could watch these first launches, the Shepard launch for instance. [For] the spot we were taken to they borrowed from some high school [or] a junior high school somewhere in Florida, the stands that they used at their baseball games which seated, probably, 12 people, maybe 24. That was the pressstand, and there were a hundred people vying for 24 seats. The broadcasters and the cameramen were expected to stand in the open.

I'm not kidding about the rattlesnakes; they were all through this [area]. As a matter of fact, it is a nature preserve and they're preserving rattlesnakes, preserving them for the arrival of the press (laughter).

And this was in Florida long before the present. If we could have recounted the snakes, we never...[laughter in reference to the 2000 Presidential election]. At any rate, we were out there in the snakes and mosquitoes. My gosh, the [mosquitoes] were thick.

But we were in this truck there. It's indicative of the kinds of facilities that we had at the beginning to the end, to tell the greatest story of mankind, of humankind, in the 20th century which was the conquest of space, from my mind, with all of the great scientific technical achievements. The one that's going to live 500 years from now in deepest minds, the date that kids are going to remember, the date that human beings escaped from their own environment into the space where they're going to be living 500 years from now. This was the way we started the coverage of that great adventure, standing out in that field of prime tobacco.

Robert Wussler: To lead into something that Julian may be able to follow up on. In the very early days, the first time that we ever did a live video or live video pictures, which then were recorded and played back off the Cape, the process that went on with anything down there was that once there was a launch, the local public affairs officers of the Air Force would write a release. Then they'd read it to Washington and Washington would say, go ahead and release it. Well, on this particular launch, launch had taken place; we recorded it, we played it back, and all of the sudden General Arnold Lehman in Washington was screaming in the phones, well, where's the release? I just saw the launch on NBC. It changed the whole philosophy.

Julian Scheer: I was still a reporter for the Shepard flight so I was not responsible for those days. But to tell you how primitive it was, Bob Gilruth was in charge of manned flight. NASA was still operating out of Langley, Virginia, and two days before the Shepard flight, he would not tell us, the reporters -- there were about 400 of us at that time -- who the first man in space would be. Gilruth had a press conference at the Cape and said, it will be either, and I'll give these to you, he said, in alphabetical order: either Glenn, Grissom or Shepard. And you can speculate, and I'll tell you tomorrow. Forty-eight hours before the launch, we found out it was going to be Alan Shepard. That's the first notification we had of who the first man in space would be. It was primitive to say the least. And there were no post-flight press conferences. There were no preflight press conferences. A couple people at NASA had gotten together with a typewriter and a old mimeograph machine and had done a press kit the night before. This is after the devastating losses we'd had with the Soviet Union.

Then, after Sputnik, to show you the tension of the times, after Sputnik, speaker McCormick, who was the Speaker of the House here, went on radio at that time and some television -- television was still pretty primitive -- and said that we faced national extinction if we did not catch up with the Soviet Union. We rushed to the launch pad, and we had the Vanguard disaster after that. The only time we really redeemed ourselves as a nation, we thought, was with the Shepard launch. One person who was totally opposed to live television of the launch was John Kennedy. Kennedy did not have a television set in the oval office. There was one on Evelyn Lincoln's desk outside of his office and members of the National Security Council and Jerry Wietzner, National Science Advisor, and, later, the President left the Oval Office and went to Evelyn's office and watched that first launch.

Joel Banow: I just want to make a point. That studio in the back of a stationwagon, is kind of symbolic, in a sense, of what, from a director's viewpoint, my viewpoint on this whole thing is. [It was] slightly narrowed in that the responsibility that I had, was to try and make the various space shots something visual, because, essentially, it was a radio story in the very, very early days.

That studio in the station wagon eventually evolved, as so many things did from a production standpoint, into a huge, air-conditioned studio, with a huge picture window overlooking the launch pad. But the space shots, even beginning with the Mercury manned shots, (though there was a remote at the Cape that was a major aspect of the production), was the space studio in New York City, where all of this was coordinated from. So with our people at the Cape, we had our people in New York City, and in the early days of Mercury, all I can remember, really, way back then, on Mercury, was that a lot of our coverage would remind one, if they were looking at it today, of an old Hollywood sci-fi B-movie. The technology we had, the way we tried to illustrate things and the only real example that sticks out in my mind, is when we had the voice of mission control, we just pulled out an old engineer's oscilloscope, which is, essentially, an electronic device with a round little faceplate on it, and you had a little jagged green lines. When somebody talked, it moved up and down. We put a label on it, and that's what we went to for the voice of mission control.

Walter Cronkite: Actually, that was really more concise and informative than the real voices (laughter).

THE FLIGHT OF JOHN GLENN

Julian Scheer: You may remember that Glenn was asked later what he was thinking about when he was on the launch pad, and the only thing he could think about he said, was the rocket was built by the lowest bidder.

John was a great favorite of the press, and the press really wanted John to be the first man in space. It was a great disappointment. Petitions were being circulated; some of us signed a petition for NASA to replace Alan Shepard, whom we didn't care much for, with Glenn.

Shepard was a pretty arrogant guy. He hated the press, and we hated him in return. Glenn was really a favorite. Glenn was really a straight arrow. The day before the launch, NASA released a picture of Glenn running on the beach, a famous picture of Glenn jogging on the beach, before jogging was really the national pastime. We had a press briefing, and they didn't tell us very much.

I went to Jack King, who was in charge of NASA public affairs at that time and said, tell me what Glenn did in addition to jogging, and he said, well, he just ran. I said he had to do more than that. Finally, in exasperation, he said, he stopped and examined a sea turtle nest.

I was covering then for all the night papers, and one was the Miami Herald. So I did the story that day, saying on the day before the launch, John Glenn ran on the beach and examined this sea turtle nest. That story ran on the front page of the Miami Herald.

So the night before the launch, I'm sitting in my room at the Holiday Inn, and the phone rings and this voice says, Julian, this is John Glenn. I thought somebody was kidding, and I said, yeah, this is John and I'm Alan Shepard.

So I realized it was John, and he says, you know, it is a federal offense to disturb a sea turtle nest (laughter) and he said, you really have a put me in a very difficult spot now. I said, well, John, it's really not that serious. He said, look, people think that John Glenn broke the law. And I said, well, what do you want me to do about it? He said, well, I think you ought to run a retraction. I said, I will.

I had a great story the next day that said, what did John Glenn do the night before he went into space? He called Julian Scheer, your reporter, and complained.

So the story ran the next day on the front page of the Miami Herald. I had, really, two exclusives in a row.

To this day, when I talk to John, he will always remind me that the first meeting we had or the first encounter with the press was an embarrassment. He still thinks about that and still thinks that I really impugned his integrity.   That's the way John was then and John is now, has not changed. He is the straightest of the straight arrows.   I want the record to say, publicly I've never said it before, he stopped; he looked, and he kept running as far as I know (laughter).

Walter Cronkite: You don't think, Julian, that it was your story that kept him from getting the democratic nomination, do you? (laughter)

Julian Scheer: Could have been, could have been.

Robert Wussler: We put up a large 20-by 30-foot screen on top of the mezzanine level for those of you who have been in the old Grand Central Station [in New York City].   It worked very effectively for us because there was an old radio show called Grand Central Terminal or something from years ago. The cut line on the radio show was "crossroads of a thousand lives" and the naked city had six million stories.

Joel Banow: They were all in Grand Central Station that day.

Robert Wussler: I always remembered that. As a result, we always put a big screen up there in the days when there were no big screen televisions so that people could get a glimpse of what these launches looked like.

James Kitchell: It was a one up [on NBC].

Walter Cronkite: They could also watch NBC coverage from a small 16-inch monitor that they had on 49th Street.

Walter Cronkite: I think that [the ticker tape parade for John Glenn], if I might suggest, tells us something about the national attention, indeed, the international attention, but particularly, the national enthusiasm in this country for this space program.

You remember the 1960's was, undoubtedly, the most turbulent decade in, perhaps, American history, including the 1860's with the Civil War. We were torn in this country by Vietnam; we were torn in this country by the civil rights struggle. And what we were seeing on television of that struggle for the first time; we were torn by the assassinations in our country. And yet, we had this one thing from Cape Canaveral we could, instead of looking down despondently, we could look up to the stars.

This nation embraced that source of satisfaction, of almost condolences to this nation for what we were going through, and knowing that on the other hand, we were making this amazing progress into space. As we made each of our successful, our first successes after those original Soviet successes and passed them so quickly, and then with the moon program, backed them off the stage entirely. The American people responded with enormous enthusiasm for the program.

Robert Wussler: And remember, this all took place at that time when there were still numerous newspapers around. I think in New York City, we still had four or five newspapers at the time. But in terms of television, there were two and a half networks. There was, NBC, CBS, and sort of a part network in ABC. There were no satellites; there was no home VCR, no Internet, no cable, no CNN, nothing like that. So in terms of video news, television news, it really came down to the three of us to get the word out. That's a lot different than the world is today.

James Kitchell: And I think it's important to note that television is a visual medium, and the two and a half or the three of us had great challenges in trying to visualize something that was going on a hundred miles up there that we couldn't see. We did have privy information relative to flight plans and so forth, but we all devised various method of trying to demonstrate what was going on in space. I've got to say that I'm very pleased to look out here in the second row and see one of the NBC producers, who was very much involved in development of visualization of space walks. He worked very closely with the Bill Baird puppets in New

York to create the simulations of the space walks. Bob Asmond is sitting right here. He's one of my associates.

Joel Banow: And you know that was, for me, the most exciting things about this was to be able to have almost ten years or more to really refine a product, a visualization as a director.

I remember the early days of Mercury when we needed to simulate so many things. For example -- the reentry; with one of my fellow directors, who was out in St. Louis at McDonnell-Douglas, we had to simulate the splashdown sequence. I remember him, we worked it out and he would try to sell me on this visualization of taking a model spacecraft on a parachute and hoisting it up on a derrick and then just letting it go and photographing it, for that part of the splash-down reentry sequence that we couldn't see until we were -

Walter Cronkite: You didn't tell me that's what you were doing. I thought that was a parachute.(laughter)

Joel Banow: Until years later, we refined the process, and, as television techniques were refined and the equipment from analog to digital, we just kept taking these same sequences and refining them over and over and over again.

Another thing, for example, how do you simulate an engine burning in space? Well, with lack of oxygen, there really isn't a great deal of flame.

I think from the very beginning, we at CBS, at least, said that we were going to try and simulate things as realistically as possible. We'd always disclaimed that it was a simulation or an animation, but I always wanted to be as scientifically accurate as possible.

So we kind of threw out the idea of taking a bunsen burner and trying to hide it behind a model of a spacecraft and have this huge flame coming out. So we kind of rendered our engine burns through 35-millimeter film animation so that an artist could render that and have control over that and, also, the way the sun was and where the light was coming from.

Robert Wussler: We burned a few studios down, too. (laughter).

Joel Banow: It was just this kind of philosophy behind, you know, of things here at CBS to try and simulate all the various things that were happening during a space mission.

Robert Wussler: You also have to remember that since there were only a few of us, Jimmy and I were competitors. At the time Jim ran the NBC space coverage, I ran the CBS space coverage. We were highly competitive and, there was great deal of stealing of each other's plans, of finding out where we were going to have cameras out in the field.   I remember Jim had a producer out in Hollywood.

At one time I got to know the producer, Jim's producer in Hollywood, very well.   And he said, are you going to put a camera at the Jet Propulsion Laboratory?   I said, oh, no, I wouldn't think of putting a camera at the Jet Propulsion Laboratory.   He said, fine, and NBC won't put one out there.

When the space shot came up, of course, we put a camera out at the Jet Propulsion Laboratory. Jimmy got me on the common phone and said, what's that camera doing at the Jet Propulsion Laboratory? I simply said on the phone, "Jimmy, I lied.".(laughter)

James Kitchell: And he wasn't the only one. (laughter)

Walter Cronkite: You know, this matter of animation -- am I permitted to tell our CBS story of animation of the moon landing?

Robert Wussler: Absolutely. I was going to tell it.

Walter Cronkite: At this stage in our history, we had Bobby Wussler's huge screen. By then the screen was even bigger, I think, in Grand Central for the moon landing, and people were really packed in there. I mean, they were packed in there, more than voters in, you know -- [laughter in reference to the 2000 Presidential Election].

Anyway, they were jammed into Grand Central for the moon landing, and we had a superb animation of the landing on the moon. These were all labeled clearly, constantly labeled as animation, and construction and so forth.

Ours was really a great picture of the lander coming down and settling down in the dust of the moon. It was timed, of course, precisely to the flight plan, so that we rolled this prepared tape, and it was to the split second of what was happening up there on the moon. And, of course we were getting reports and the audio report from the landing as to how it was along in little bits and pieces. If there was a little space, I would try to fill in what the situation was at the moment.

Well, we've got that marvelous lander down on the moon, flying moon-dust around it, the lander down, and I'm saying -- you know, I had as long to prepare that as NASA did, for heaven sakes, and suddenly, I'm speechless. "Oh boy, oh boy, we've got a man on the moon," except that it was three and a half seconds, I think, before the lander actually got on to the moon because they had last minute glitch, and indeed, they darn near lost the lander if Neil Armstrong hadn't done a superb job of lifting the lander up and moving it a few feet away from a rock that would have tipped them over. He had to move and that took another three and a half or four seconds.

During that time, the crowd in Grand Central Station, and as far as I know, and around the world was listening to CBS and celebrating man's landing on the moon. For three and a half seconds, the CBS future, Bob Wussler's and certainly mine, hung in the balance. If Neil Armstrong had not gotten that lander down, at that moment, we would have been in deep trouble.

Joel Banow: That was always the trap because my script for that mission was the actual mission flight plan. I used that to put all my queues and so forth and we did, precisely, have all the information as to how long it was going to take, from engine burn to touchdown and, of course, as Walter just said, we didn't take into account that there may be a few rocks in the way or something. There it was; it came down. We're sitting there, and we're still hearing, you know, 20 feet down to the left, two to the right, and then, oh, we're down already. But, through the luxury of video-tape, and the fact that we had to go back on the air later on and recap the landing up on the moon, we were able to go in and sync up the audio. When we had them land, they landed exactly on the right time, with exactly the right word from that point on throughout our coverage.

Walter Cronkite: I'd like to make it clear that this was many, many, years ago. Today, nothing like that happens, of course (laughter).

James Kitchell: I think we have to give some credit to Julian and his organization who finally realized that television was a major part of what was going on with the space program and the public reaction to it. They did then manage to squeeze in a television camera early on, and then, some fairly sophisticated television cameras aboard the later flights. We thank you for that, Julian, and the public thanks you for that because we couldn't all be right.

Julian Scheer: One of the interesting things on that landing when Neil Armstrong stepped out on the porch for the first time, he forgot about the camera and reached back in and pulled the lanyard that dropped the camera down so you could see the first step. But for a few seconds there, we didn't think we would have a picture at all. Then Buzz [Aldrin] kept saying, the camera, the camera.

Walter Cronkite: Buzz would. (laughter)

RECOLLECTIONS ON GEMINI

Julian Scheer: This first spacewalk with Ed White. We had planned for a long time for Ed White to stand up in the spacecraft and not to get out, but Alexei Leonov three months ahead of Gemini Four, did a spacewalk for about four or five minutes. So the astronauts really lobbied vigorously to have Ed White get out on a tether and do a longer space walk. When they came to headquarters here, in Washington, it was really very controversial.

We were split down the middle. Dr. Drydon, who was the number-two man at NASA, was strongly opposed to it; Jim Webb finally supported it. The Apollo office was divided and the decision was made to go ahead and do it.

The hand-held device that you see had been developed by Ed Gibbons, who was an astronaut who was later killed in an automobile accident. Ed White and the crews has been practicing for months without general knowledge even within the program. They had a hanger down in Houston, and they would use the maneuvering device and skid around the floor. So when the request for an extra vehicular activity, the space walk mission, came to Washington, there was a great deal of skepticism at that time; but it finally went forward.

It was timed to be exactly twice as long as Leonov's. So, that was considered a great space accomplishment.

One little sideline to this, several weeks after the launch Ed White called and said, do you mind if I write a letter to President Johnson? And I said, you know you can write to anybody you want to. We had the ability to be open be free and open with our crews. It seemed his father had been in a United Nations peacekeeping operation. His father was a general in the Air Force and had a United Nation's flag and, Ed, without telling anyone, had put it in his pocket on the inside of his leg and had that flag with him in his space walk. He sent it to President Johnson, who turned it over to Ambassador Goldburg, who then turned it over to the United Nations, and it's hanging in the United Nations today. Ed White, of course, was later lost in the 204 fire.

As far as the Gemini Eight, our most anxious moment up to that time, as far as space was concerned, we thought that the 60 revolutions per minute was sort of at the critical point as far as black out and orientation was concerned.

I had left the control room, and I got a call saying that the spacecraft had separated from the Agena and was spinning out of control; and that the astronauts would be lost in space; and that they had been saying things that we would have to make a judgment.

I didn't want the family to hear [their final words]. I thought, you really had Dave Scott and Neil Armstrong uttering their final words. I didn't think it was really fair to the families to hear that. They each had little microphones in the home so they could hear the out-of-ground transmission.

After the spacecraft righted itself, when it was very clear that everything was all right, I listened to the tapes and there was absolutely nothing on there that would even tell you that they were in a dangerous situation. Neil was completely calm all the way through, and so was Dave Scott. They were completely in control -- no anxiety whatsoever in their voices.

We released that tape later; it's the only tape we've ever withheld. It was a bad judgment on my part, but at the moment, I was told we that would be hearing the final words of two crewmen in space. Now, we had lots of contingency plans that we carried with us all the time just in case something like that happened. We had contingency plans and notification of next of kin for launch pad, lift off, lift off before injection, in orbit, in transit to the moon, on the lunar surface, coming back and in splashdown. So what we really kicked in at the time of Gemini Eight was the contingency of astronauts being lost in orbit.

Fortunately, it didn't happen, but the record will show that for the one and only time in the ten years I was at NASA, we withheld a tape. One other tape we did withhold, I must say, we did have about four or five seconds of the astronauts just before they died in the fire, and I made the judgment that the tapes would never be released and has not been released or has never been heard by anybody expect the technical people. But everything that happened in orbit from all the NASA missions -- I went to work for NASA at the last Mercury flight, from Mercury all the way through Gemini and through all the Apollo missions -- were released in real-time without editing. But, you know, I had to live for a long time with knowing that we withheld a tape that was unnecessary.

Walter Cronkite: I made a little speech a moment ago about how the American public was so excited about space flight, but I've got the reverse side of the mirrored report to you on Gemini Eight.

I was at our studio at Capcom just a little bit before our evening news program, that they lost control or that they didn't lose control, that they were unable to regain control. They were actually out of control in space. To us, it appeared that the chances of their recovery were very slim as Julian says.

But we went on the air immediately, interrupted the program in progress, which was [in the] late afternoon / late early evening and started reporting this story. I was deeply afraid that we were going to be reporting the loss of our first astronauts.

But our stations around the country were overwhelmed by telephone calls from people complaining that we had interrupted the program in progress. The program in progress was a futuristic serial drama called "Lost in Space".(laughter)

James Kitchell: And our other competitor, ABC chose not to go on the air immediately, because they were premiering "Batman" that night. (laughter)

But that was kind of a tough night for all of us because once we all had the benefit of having copies of the flight plan, as soon as we got some communication from the spacecraft that they had activated the RCS, the reaction control system, we knew that they were going to have to land within one orbit. We did not know where.

As it turned out, it was in the north Pacific. It was probably, if I recall, something like 350-miles from the nearest ship. We had to stay on the air until we knew that the astronauts were safe. There was no communications with them. And we all scrambled, trying to figure out, how are we going to explain this to the audience?

One of the things that we had developed at NBC was Frank McGhee's ocean on the studio floor which was a blue floor, and we had models of ships and so forth. That was the sum total of our visual activity for about three and a half hours.

Walter Cronkite: The big trouble was, remember Jim, they landed in the Pacific instead of the Atlantic. That's why the ships were so far away.

James Kitchell: That's right.

Walter Cronkite: We had emergency vessels out there but not the primary recovery boats.

James Kitchell: The primary recovery force was in the Atlantic, but they came down. They had to, under mission rules, did have to make a landing, and they landed in an area where they were no contingency forces. We had to try to cover for that until we did get some positive word that they were safe.   That was one of challenges that existed for television throughout all of this. We've talked about simulations and so forth but trying to deal with the unexpected was, at times, a strain.

Joel Banow: It was, and that's where, you know, the graphics departments and the production departments really have to scramble. With the Agena docking part of the mission, we had miniature models, and we had full size mock-ups. Here we were, going through the simulation; and we docked when they docked because you didn't see it live. So we had to do it ourselves.

Walter Cronkite: This is spoken like a director, not thinking about the poor guy on the camera, who has to fill the time while they're thinking of what they can put up on the screen.

James Kitchell: Admirably, I might add, Walter.

Julian Scheer: Well, we had to bring -- because they used the thrusters to control the roll -- we had to bring them in right away. We had the oceans of the world pretty well covered, but, as I remember, the nearest vessel which was a destroyer, was about 350-miles away.   So they bounced around on the water for quite a while before they were picked up, and we finally got them to Guam. And Wally Schirra was in Hawaii and, you may remember, flew from Hawaii to Guam and met them; we brought them home from Guam. It was really a pretty tough period of time.

Walter Cronkite: One of the worst things that we faced at all times, as far as I was concerned, at any rate, we had tons of material to fill in on a normal flight, but it was the possibility of a loss of an astronaut or a crew.

It was a terrible burden that just might happen. The worst experience we had, I think, was a totally unnecessary one when [M. Scott] Carpenter, in the Mercury program, he was, I think, the third flight, if I remember, third or fourth. Any way, at any rate, Carpenter came through the blackout as you all know about, I suppose, the blackout as the space vehicle reenters the earth's atmosphere -- a temperature of 3,000 degrees on the nose of the spacecraft -- and during that period of time, about two to three minutes in there, there is no real communication, certainly no voice communication at that time.

There was always this desperate wait for the craft to get through the blackout and again, to communicate again with mission control. There's this exaltation always in mission control and among us, who were reporting it, and, I think, the people who began to realize what it was all about: this exaltation that the craft had survived the reentry into the atmosphere, this very crucial moment.

Well, with Carpenter, he entered the blackout, and that's the last we, being informed by mission control, heard for 47 minutes. They told us they did not have communication with Carpenter; we've got brief reports of that. What they did not tell us what that they had biological reports that he was alive in the craft. But they didn't tell us that, didn't communicate that to us. As far as we knew, he was lost in space. I mean, he was lost; he never came of the blackout successfully. It was only because of a miscommunication from mission control, not Julian, I don't mean to point to him, he had nothing to do with this (laughter). He had nothing to do with it except he was the boss of the man that had something to do with it; not exactly, I suppose.

James Kitchell: That was pre-Julian.

Julian Scheer: Well, Scott [Carpenter] missed the footprint almost entirely, I mean. I was a reporter at that time. We thought that we had lost him. He was way, way off, and it was always considered pilot error.

James Kitchell: Well, he had done a manual reentry or manual retrofire. I had some anxious moments because the evening before his launch, he had slipped off the Cape and done a drop in at an NBC beach party.

Walter Cronkite: But the problem was, I think, you folks understand the problem, here we were, on the air, faces hanging out, nothing to show from these guys at the director's office, how do we handle this matter? We didn't want to say that it appears that Carpenter been lost, we were ducking every one of those words that would have alerted the public to the probability, as a matter of fact, rather than possibility, that he had been lost.

I want to tell you, vamping around that series of circumstances for 47 minutes was about as exhausting an emotional experience as I ever went through on the air.

THE APOLLO ONE (Apollo 204) FIRE

The official announcement came from NASA protocol officer Jay Veeham.

(Officer Veeham): Astronauts Virgil I. Grissom, Edward White and Roger Chaffee were killed tonight in a flash fire during a test of the Apollo Saturn 204 vehicle at Cape Kennedy Air Force station. The fire occurred while the astronauts were in a spacecraft during the countdown of a simulated flight test. The accident occurred at 6:31 p.m. Eastern Standard time at T-minus ten minutes prior to the planned simulated liftoff. The spacecraft is located 218 feet above the launch pad and was mated to the upgraded Saturn 1-B launch vehicle. Hatches on the space-craft were closed. Emergency crews encountered dense smoke in remov-ing the hatches.

Robert Wussler: We were all convinced that following that fire that the original date of putting a man on the lunar surface by the end of the decade would not happen. But NASA did a terrific job, pulled them-selves together, did the appropriate post-fire tribunals that needed to be put together, and within two and a half years of that fire, we were able to put a man on the moon. Incidentally Ed White, whom you saw earlier, the first man to walk in space, who was killed in the fire in January of '67, was extremely popular among the media. I don't know what Julian's feel-ing is on this is, but at least among the media, he was sort of the odds on favorite to have been the first man on the moon.

Julian Scheer: I think that's right. It was the lowest moment, I guess, in the program for all of us, I think for the American people, the people in the space program and for the media. It was an exceptional crew.

Roger Chaffee always outstanding; Gus Grissom was a favorite of a lot of us. He was really a very hard-nosed, tough guy, had a rough career in NASA, losing a spacecraft in Mercury.

Ed White really was an exceptional guy. He was at the top of his class at West Point. He was a NCAA 440 runner; his father was a General. He was very handsome, very skilled, and had done the first space walk. I think he really did have -- he was one of two or three members of the astronaut group that really stood out as exceptional in a lot of ways.

It was really a very, very, low moment for us. And we went through long Congressional hearings. We went through a long investigation. We moved the spacecraft immediately to Langley, Virginia and we put together a team with Frank Borman and Max Faget and Floyd Thompson. They did a thorough investigation. We redesigned the spacecraft, changed the environmental control system entirely, changed a lot of the electrical stuff. We had several people in the program leave the program; several people in the contract group were dismissed, and we had a tremendous internal and external shake up. But the fixes were made.

The program had a lot of momentum. We had strong support from Congress. We announced just five months later that we were going back on schedule. It hardly caused a ripple. I think a lot of us thought in the first 24 hours that the space program would end at that point.

As Walter mentioned earlier, the question of losing astronauts had always been an issue, and whenever we came close there was always a certain amount of outcry in the public that maybe going to the moon is not worth the loss of life. It was really epitomized by the 204 accident, but I think, while we did everything in NASA we could to make missions as safe as possible, I think we had a final conclusion, always, that an accident could occur and was likely to occur. Of course, the Challenger accident later, unfortunately, proved that wisdom to be true.

The tool for an accident, looking back on it, was really primarily a design fault. Gus Grissom had hated that spacecraft. He'd complained about a number glitches not related at all to the fire. He kept talking about for weeks that he was going to hang a lemon on the outside the spacecraft. As a matter of fact, the day that the fire occurred, he had, indeed, done that. It was a very prophetic thing.

When Betty Grissom was told about the fire, the first comment she made was, did Gus get a chance to hang the lemon on the spacecraft? And my answer was, he had. Gus had a very ominous feeling about the spacecraft. None of us thought the design was incorrect, but it was a very radical design change which made all the future Apollo spacecraft, I think, acceptable and far better.

The changes were very radical in several respects including the door, which was really the problem. The pressure in the cabin made the door impossible to open for an escape. We changed the design completely after that.

THE MOON YEARS

Julian Scheer: I found some numbers. We had 3500 reporters at the Cape and 1500 at Houston at the launch [Apollo 11]. CBS had 244 people assigned to it. Bob, did you realize that?

Robert Wussler: And Jim had only 147.

James Kitchell: I saved a bunch of money. (laughter)

Julian Scheer: I had 118 reporters from Japan, and 82 from England, and 81 from Italy, and 53 from France, and 52 from Argentina. We had reporters from Canada, and Spain, Brazil, Mexico, Switzerland, Angola, Chile, India, and so on. It was an incredible thing. One Yugoslav newspaper I found offered an $800 prize --why $800 I don't know -- for anyone who could guess what Neil Armstrong would say when he landed on the moon.

You know, I had a strange reaction [to the Apollo 11 landing]. I was in the control room, and as soon as Buzz got out, they erected the flag. The flag staff would not go in very far and the flag would not unfurl. We practiced a lot. There's no atmosphere so we had to design a flag to just drop down. When they left the lunar surface, obviously, we knew it would blow over.

My reaction was, now that you've done it, get the hell back in the spacecraft and come home. I was not at all enthralled with the idea that they'd spend a lot of time on the lunar surface. I mean, I was so nervous and so excited at the same time. It really was an exhilarating moment. Everybody had the same reaction. I really wanted them to --my thought pattern went, you've done it; you've been there. It doesn't make any difference whether you're there for two minutes. Just come home. I was concerned about whether we would be able to successfully liftoff the moon. So all the time they were on the surface, I kept wondering if everything would go right.

I was ready for them to come home right away. I think it was, for a group of people, and we had 30,000 people at NASA and 200,000 people in the contractor group, and thousands of people on college campuses, and everyone had a role of some kind and had worked for a decade.

There was a collective excitement, I think, that ran through this great mass of population. Everybody felt a part of it; everybody felt they had a part. Everybody worked toward this goal. It was very unifying, almost religious experience. As Walter mentioned earlier, in 1968 just six months before, we had two assassinations and the Vietnam War. You had kids staring at the sun and going blind. You had a lot of acid and other drugs out there.

You had two cultures going on in the country. You had a bunch of crewcuts -- you heard Charlie Duke on the capcom -- you had all the crewcuts at NASA and you had people with very long hair. We had the long hairs and the short hairs (laughter as Martin Collins pulls on his long hair).

I mean, it was quite an interesting thing, and we lived in sort of an interesting, peaceful detente. While you would expect people who were engineers and scientists and technologists to be at war with this other group of people who had an entirely different interest and almost an entirely different culture. There was something unifying about landing on the moon and about being first. I must say that we were very conscious throughout all these years, particularly though the years of the Cold War, that we were in a race.

Two years ago I met Alexei Leonov and spent a day with him in New York. Leonov was the second Russian in space and the first to do a space walk. I had met him in Greece a number of years earlier when he was a cosmonaut. I asked him about the race. He said, well, we did have a race. We thought that we were going to win, and I really thought I was going to be the first man on the moon.

He said, when Frank Borman and his crew went around the moon on Apollo 8, we gave up. We decided then that there was no way we could catch up. But he said, until Apollo 8, we really thought we would be on the moon before you, and your crew denied me, personally, the opportunity to be the first man on the moon, which I thought was kind of interesting.

Robert Wussler: And, Julian, one quick point. Apollo 11 took place during July of 1969. Not only was Vietnam at its high point, but in a 30-day period, Apollo 11, Chappaquiddick, the Manson murders and the Woodstock [music festival] all took place.

FEELINGS & WORK ON APOLLO 13

Joel Banow: One of the things I remember is missing a wonderful steak dinner (laughter).

We had taken a break, and we were off the air. I'd gone across the street to a CBS watering hole, a restaurant. I'd just ordered a wonderful steak dinner for my meal break. I got a call from this guy over here with his, "we've got a problem." And I never did get to have that steak dinner as I tore back across the street to the production center.

It was something, where again, television news showed what it could do in adjusting to a major change in something that we were doing which was the coverage of another Apollo. We got on the air and told the story, and we adjusted to what we had to do, changing some of the things we had preplanned. As I had mentioned earlier about the Agena docking, once they had that problem, we quickly had to change our models. We had to undock everything; we couldn't have them together in here. Everything that was an Apollo command service module we had to change because there was an explosion. So we quickly had to have our graphics department render the side of the spacecraft so it would look like something happened to it. As soon as we got accurate information, we kind of rebuilt our models and made them look like what we thought was the problem so that we were kind of in sync with what was going on.

Walter Cronkite: The biggest problem we had, Joel, of course, was that we didn't know, even as the astronauts did not know, nor did ground control know what had happened. All we knew was that they were in serious danger, and we thought we had probably lost them. We didn't know how serious the matter was or what it was. So you didn't have anything to illustrate, and we didn't have anything to talk about.

Joel Banow: What I actually meant was, when we did find out what the problem was after they undocked and were able to view and then sent back their descriptions, as you can see there [photo of Apollo 13 on screen], we then quickly adjusted, as I just told you, to what they said they saw. I just kind of compressed the time there.

James Kitchell: I think the [Apollo] 13 episode brought us all back to some reality. We had gone through the fire. We had recovered from that; we had gone forward. We had landed on the moon twice. The public was becoming rather blasé about it, and the [Apollo] 13 episode brought us all back to reality.

FINAL REFLECTIONS

[…on visiting Mars]

Walter Cronkite: When? I'll take a quick shot at it. It's a long time off. We've got to do a lot of robotic exploration of Mars before we can even think of getting a man up there. We've had two unsuccessful flights that were meant to land on Mars and do some further work, like that marvelous Surveyor did up there before. We've had two failures. The next landing on Mars are now scheduled for, at the earliest, 2003, I think, maybe later than that now. We're a long way from putting a man up there. We've got a lot of work to do with robotics. Robotics are a wonderful way to explore the universe. It's not mandatory at all that we get man out there beyond the moon at this point. We even probably will begin to use the moon as a launching site for deep space exploration before we get any men or women out that far.

[...on going up into space?]

Walter Cronkite: I think about it all the time today. I'd be ready to go tomorrow if they'd ask me and if I could pass the physical which I probably couldn't. Of course, I can't imagine a greater adventure than going into space. When people tell me they wouldn't think of it, I dismiss then from my group immediately because I don't want anybody that unimaginative around me (laughter).

The only thing that I think I would have a problem with, quite honestly, if I had a chance to go up on the shuttle, for instance, is knowing my own temperament, I would see the glass not as half full but as half empty because I wasn't going to the moon.

That's where I'd like to go: to see this Earth, to see this wonderful Earth of ours from that orb out there, the moon. That picture you may have seen of the Earth rising with the horizon of the moon, I think, is the most graphic and wonderful picture I've ever seen in my life. As the astronauts themselves said out there, this is the only colorful planet in the entire universe as far as we can see it with the naked eye and even with our magnificent telescopes. This is the only bit of color out there.

You know, when we look at that Earth of ours out there like that, and I think they got a religion of their own when they saw it, and we should all, I think, get a certain religion to understand that we're on a fairly precious planet, a precious, precious planet with life on it; life with all that means.

And to think that we can't along on that little planet of ours and believe the only way we can solve our problems are to shoot each other. I think we must come to a new philosophy. Once we can see our Earth from a distance and understand how vital it is, we should appreciate what we have and try to save it

SpaceBusiness.com Order Form

Fax orders: +1 (301) 718-1837

Telephone orders: +1 (703) 524-2766.
Have your VISA, AMEX, MasterCard ready.

On-line orders:
Visit the bookstore at http://www.spacebusiness.com

Postal orders: SpaceBusiness.com
P.O. Box 5752 Bethesda, MD 20824-5752 USA

Please check the product you want to order...

___Please send ____ copies of *"In Their Own Words"* at $14.95 each

___ Please sign me up for a one-year subscription (4-quarterly issues) of *Quest: The History of Spaceflight* at $29.95.

Name: _____

Company Name: _____

Address: _____

_____

City _____ State ____ Zip _____

Country _____

Email or Tel #: _____

Sales Tax:
Please add 5.0% for books shipped to Maryland addresses.

Shipping:

U.S. Post Office: $2.00 for the first book and 75 cents for each additional copy. (Can take two to three weeks)

UPS: $3.00 for the first book and 75 cents for each additional copy (Delivery time is three to five days)

International Air Mail: $4.00 per book

Payment:

__ Check Enclosed   __ VISA   __ Mastercard   __ AMEX

Card Number: _____

Name on Card: _____ Exp Date: ___ / ___